WHAT ... S ARE MADE OF

WHAT
STARS
ARE MADE OF

The Life of
Cecilia Payne-Gaposchkin

DONOVAN MOORE

Foreword by Jocelyn Bell Burnell

Harvard University Press

Cambridge, Massachusetts
London, England
2020

Library of Congress Cataloging-in-Publication Data

Names: Moore, Donovan, author.
Title: What stars are made of : the life of Cecilia Payne-Gaposchkin /
Donovan Moore.
Description: Cambridge, Massachusetts : Harvard University Press, 2020.
Identifiers: LCCN 2019031057 | ISBN 9780674237377 (cloth)
Subjects: LCSH: Payne-Gaposchkin, Cecilia, 1900–1979. | Astronomers—
United States—Biography. | Women astronomers—United States—Biography.
Classification: LCC QB36.G372 M66 2020 | DDC 520.62 [B]—dc23
LC record available at https://lccn.loc.gov/2019031057

Frontispiece: *Cecilia Payne-Gaposchkin*, by Patricia Watwood, oil on linen,
47 x 38 inches, 2001. Collection of the Harvard Art Museums.
Copyright by Patricia Watwood.

For Brendan and Hana

Contents

PART III. DISCOVERY
Harvard, 1923–1979

Foreword

JOCELYN BELL BURNELL

When I was an undergraduate at the University of Glasgow in the 1960s, everyone had assigned seats in the lecture theater (to aid in taking attendance). Inevitably, the women were placed in the front row. When we entered the theater we were often greeted by whistles, cat calls, the thumping of desks, and the stamping of feet by the male students. Forty years earlier, in Ernest Rutherford's classes at the University of Cambridge, Cecilia Payne also sat in the front row, and she faced similar "attention." The difference was that in my day, lecturers did not (normally) encourage this behavior the way Rutherford did. That indicates some progress, I suppose!

Einstein had praised the mathematician Emmy Noether as "the most significant, creative mathematical genius thus far produced," but my (male) undergraduate math lecturer claimed that Noether had misinterpreted the subject. He said it with a viciousness, a visceral anger, that made the remark memorable for me, the only woman in the class. When people talk about women in science, there can be strong, very human undercurrents that perhaps we do not acknowledge enough. The upsetting of the old order can be distressing; but when a woman does it, it can be too much for some.

I'm lucky to work in the United Kingdom, a country where women now have more freedom. But there are still countries where women are much more restricted—impoverishing those countries, I judge. And when Cecilia Payne was growing up in England, women's roles were also much more limited.

Girls and boys were given very different educations, and Cecilia struggled with the path through life mapped out for girls. Clearly highly intelligent, she challenged herself, as well as those who arranged the education of English girls. This attitude resonated with me, as I grew up in a small town in Northern Ireland where "everyone knew" that girls were only going to get married and be homemakers. They did not need the same education as boys, who would be in the work force and therefore of economic value.

Fortunately for me, my parents were wiser. They were determined that their daughters would have as good an education as their son, and they fought for me to have a place in the science class with the boys rather than in the cookery and needlework class with the girls. When I came top of that science class in the end-of-term exam, the teacher said nothing to me. Instead, he lambasted the boys for allowing a girl to beat them! From quite an early age, as a girl I was in a small minority. Even subsequently in a girl's school, I was one of only two in my year who studied physics all the way through.

I was lucky in that I knew before I left school what I wanted to be— I wanted to be a radio astronomer. This helped me push through barriers and quietly challenge assumptions. Cecilia Payne-Gaposchkin had to work hard to be able to do any science at all, and she didn't identify astrophysics as her chosen subject until that dramatic evening when, as a new undergraduate in Cambridge, she heard Arthur Stanley Eddington lecture in Trinity College about the eclipse expedition that had helped establish Einstein's theory of relativity.

I admire her courage when, just a few years later, she set off across the Atlantic to work in a new place and a strange country. The Harvard College Observatory had been employing a number of very able women (called "computers"). These women welcomed the opportunity to work, but they were very poorly paid and were not expected to advance. Nevertheless, they laid the groundwork for major advances in astronomy. Cecilia, as a graduate student with her own funding, was in a somewhat different category; she had to navigate her own path between personalities, always aware of what was expected of a young woman.

It must have been quite distressing for her as a PhD student to realize that when she applied the physics she had learned in Rutherford's Cavendish Laboratory in Cambridge, she obtained results (concerning the abundance of hydrogen in stars) that were seriously in conflict with what was held to be the case by all the senior astronomers, including Eddington (whom she seems to have admired above many). She was pinioned at the point where astronomy was meeting physics. I was luckier in that the validity of my "crazy" result—pulsating radio sources, or pulsars—was recognized much more quickly. I identified the first four in a six-week period, and acceptance by the community came rapidly.

Payne-Gaposchkin managed to stay at the Harvard Observatory (albeit mostly in very low-paid positions) all her working life, including through marriage and rearing a family, whereas I moved around the country with my husband and struggled to keep working while raising a child. Everybody knew, I was told, that if mothers worked, the children would become delinquent. Consequently, there were very few places where the children of mothers with jobs could be well cared for; so for many years I worked part-time. However, I was able to stay involved in the science through jobs in radio astronomy, millimeter-wave astronomy, and infrared, x-ray, and gamma-ray astronomy. The posts I held were rarely research positions, but I got to try my hand at administration, scientific support work, lecturing, public outreach, management, and leadership.

For much of her career, Payne-Gaposchkin stayed in the same line of astronomy (stellar spectra) which she had so successfully exploited for her PhD, but she was not conservative in her approach. One of the remarkable things about her was that in her later life, as new techniques (including the use of satellites) were developed, she applied her understanding of stellar spectra to the new wavelengths that became available. What an adventurous and courageous woman!

There is one other dimension to Cecilia Payne-Gaposchkin that I have never seen discussed. It is of interest to me, as it forms another link with my experience. I have been a Quaker since birth and am an active member of that church. In the 1970s, while casually browsing through a membership list for British Quakers, I was amazed to discover Payne-Gaposchkin's name. She had applied for membership in the London area of the Religious Society of Friends (Quakers) just as she was finishing her undergraduate studies. In her letter of application, dated June 4, 1923, she writes that Church of England services have ceased to have meaning for her and that she has recently attended Quaker meetings in several towns and cities. She joined the London Quakers and remained a member there for the rest of her life, even though her contact was intermittent.

I do not know how, or why, this attraction to Quakerism came about, but most likely it had to do with Eddington's influence. He was an active Quaker, from an old Quaker family, and was a member of the Cambridge meeting. She had been working with him and greatly respected his judgment.

Quakers believe that one does not have to have an intermediary such as a priest, minister, or holy man to mediate between God (by whatever name) and humankind, but that individuals can experience directly the presence of God. Cecilia Payne-Gaposchkin wrote (maybe in a different context, or maybe not):

> There are those—and I am one of them—who rebel at having to deal with an intermediary. They want to go to the fountain-head . . .

to be in direct touch with the fountain-head whether you call it God or the Universe.

Near the end of her life she wrote about what had enabled her to be so successful:

I have reached a height that I should never, in my wildest dreams, have predicted 50 years ago. It has been a case of survival, not of the fittest, but of the most doggedly persistent.

Yes, Cecilia; I know!

Author's Note

In the spring of 2015, as I was searching for a book idea, a friend sent me the materials for a course he was auditing at Princeton, entitled "The Universe." As I was leafing through the presentation, one of the pages gave me pause. It had three photographs, with no names. I recognized the two men, Aristotle and Newton; but who was the woman, literally on the same page as these great men of science?

I couldn't tell much about her from the presentation, but I did find her name: Cecilia Payne-Gaposchkin. Curious, I began to search for more information.

First, there was that photograph. It wasn't a picture of Cecilia herself; it was of her oil portrait. Patricia Watwood, an American portraitist, had painted it from a collection of twenty-five separate photos. She had based it on *The Astronomer*, a 1668 Vermeer painting that hangs today in the Louvre. The portrait was commissioned in 2002 by Dudley Robert Herschbach, a Nobel Prize–winning professor of chemistry at Harvard, and his wife, Georgene Botyos Herschbach. For years, Professor Herschbach had argued that the Harvard faculty should include more women. And so should the portraits hanging in university buildings. "Affirmative action for portraits," Herschbach said at the time.[1]

The Faculty Room at Harvard's University Hall prominently features the portrait of Abbot Lawrence Lowell, president of the university from 1909 to 1933. In 1928, he had decreed that no woman should be granted a teaching appointment at Harvard. Cecilia died in 1979, never knowing that her portrait would later hang on the very same wall, about thirty feet from Lowell's.

At the dedication ceremony for the portrait, Jeremy Knowles, dean of the Harvard Faculty of Arts and Sciences, quoted the words of an undergraduate who had just learned about Cecilia's discovery: "Every high school student knows that Newton discovered gravity, that Darwin discovered evolution, even that Einstein discovered relativity. But when it comes to the composition of our universe, the textbooks simply say that the most prevalent element in the universe is hydrogen. And no one ever wonders how we know."[2]

As I continued my research, I learned how difficult it was to be a woman of ambition in early twentieth-century England, much less a woman wishing to be a scientist. Battling a society that did not know what to make of a determined young schoolgirl, putting up with derision as the only woman in a college physics lecture hall, facing skepticism as a graduate student when her computations contradicted the learned men of the astronomical community, spiriting her husband-to-be out of Nazi Germany— Cecilia had to overcome numerous obstacles, both scientific and personal.

Ultimately, she did what every scientist yearns to do—discover. In Cecilia's case, her discovery was one of the most fundamental breakthroughs in scientific history: determining the atomic composition of stars. But the odds against her doing so were daunting. How did she do it? I asked myself.

I found out. First of all, she was brilliant. From her earliest days, she had displayed a relentless curiosity. About everything: stars, for sure; but also music, literature, art, politics, cooking, sewing, soap-making, you name it.

Maria Mitchell (*seated*) inside the dome of the Vassar College Observatory, with her student, Mary Watson Whitney, ca. 1877.

Maria Mitchell's first group of six astronomy students, known as the Hexagon. Mary Whitney is seated in the center.

Coming to America was also key. She secured a job at the Harvard College Observatory when there was no work for her to be found in England.

She also could not help herself; the hunt was intoxicating. There were times, nights mostly, when she simply could not stop working. She surely felt what the English paleontologist Richard Fortey captured so eloquently: "The excitement of discovery cannot be bought, or faked. . . . It is an emotion which must have developed from mankind's earliest days as a conscious animal, similar to the feeling when prey had successfully been stalked. . . . [It] provokes a whoop of enthusiasm that can banish frozen fingers from consideration, and make a long day too short."[3]

Hers was a mind like few others; but it was also a mind that belonged to a woman, which meant that brainpower alone was not enough. The battles she had to win on earth were every bit as big as what the universe would throw at her.

And she fought these battles alone. There were not many women scientists in the world at the time, and very few women astronomers. Cecilia's only true role models were Maria Mitchell, who had been appointed professor of astronomy at Vassar in 1865, and Mary Whitney, who succeeded Mitchell and served as director of the Vassar Observatory until she retired in 1915.

The ranks were thin. There were no student colleagues she could lean on, no hashtags to identify women in similar situations to draw support from. With an extraordinary discovery in hand that signaled real trouble for established theories, Cecilia Payne was told that her conclusions were wrong by the very man who four years later would show her to be correct. He got the credit. At the time.

WHAT STARS ARE MADE OF

Prologue

Something was wrong.

Night after night in the winter of 1924, Cecilia Payne sat in her small office, with barely enough space for a desk, at the Harvard College Observatory. Prominent on that desk was an eight-inch-diameter ashtray "piled three inches deep with ashes and butts."[1] She would squeeze her eyes and stub out the last of the day's pack of cigarettes. With money tight, and growing tighter, she would soon be forced to pawn her inherited violin. "Often I was in a state of exhaustion and despair," she wrote later, "working all day and late into the night."[2]

Cecilia spent hours staring at photographs that had been made at the observatory. The observatory's telescope directed light from an array of stars into a spectrograph. The spectrograph then split the starlight into its component colors. A sophisticated camera preserved the various wavelengths in the form of a spectrogram, which consisted of black lines on clear glass plates. (Color film did exist at the time, but monochrome provided much clearer images.) The technique was, for its time, remarkably sensitive. "Even in Cecilia's day, spectrographs could resolve something like one hundred million wavelengths," says the astronomer Virginia Trimble, "while your eye can resolve maybe a thousand."[3]

The spectrum of every star, from the hottest to the coolest, was displayed in the same way: a ribbon of black lines representing a red-to-violet rainbow. Depending on the temperature of the star, some lines were so weak as to be barely discernible, while others were strong and clear. But whether the lines were faint or bold, a particular chemical element in a particular ionized state always produced the same pattern, always left the same "spectral fingerprints."[4]

To Cecilia's trained eye—squinting as she carefully slid a jeweler's loupe around the plate—the charcoal-colored ribbons held ancient secrets. From the earliest moments of recorded history, humans had been looking up, wondering about the composition of the cosmos.

By the nineteenth century, astronomers, still looking up but now with the aid of telescopes, had identified many planets, stars, and comets. Based on the principle of uniformity in nature, astronomers assumed that all heavenly bodies were composed of the same elements found on earth, and in roughly the same proportion.

It was in the mid-1800s that astronomers were shown a different way to study the sky. Two German scientists—Gustav Kirchoff and Robert Bunsen—managed to correlate chemical elements with spectral patterns. Each element, when heated up, produced a unique set of lines. In time, astronomers would put Kirchoff and Bunsen's technique to use. They reasoned that they could use telescopes to direct starlight through a prism, which spread the light into a spectrum of wavelengths. If the spectrum were then recorded on a glass photographic plate, it could be studied. It was the birth of astrophysics.

By looking down—at those prism-spread wavelengths of starlight—Cecilia was one of the first to do what astronomers had tried to do by looking up. It was as if all that starlight, preserved on thousands of glass plates in the Harvard Observatory's catacombs, represented a quarter million jigsaw puzzle pieces waiting for the right person to fit them together.

Unbundled and captured by the spectrograph, the light was now a coded message. To a skilled cryptographer, the pattern of lines revealed the precise composition of the star from which it came. And Cecilia, trained in astrophysics, was very skilled. The spectrograms, however, were not cooperating. What the messages were telling her did not conform to existing theories. Could the entire world's astronomical community be wrong? Or was she mistaken?

She was not accustomed to making errors. She was a twenty-four-year-old graduate student, working on a thesis for a PhD in astronomy. She had grown up in London, and had just graduated from the University of Cambridge. Her major was physics. She had been trained by Nobel laureates in Cambridge's famed Cavendish Laboratory, an unforgiving place where measurement—the end result of observations made with rigor and precision—was prized above all else.

As she studied the plates, she relied on her training in astronomy and physics to recognize the patterns that indicated the presence of silicon, of magnesium, of aluminum, iron, helium, hydrogen. What she found was making her suspicious of the widely held belief about what stars were made of. Hydrogen was showing up in much larger amounts than it was supposed to. That meant that stars were not just heated-up earths. They were actually composed mostly of hydrogen. And not just a little more hydrogen than on earth—a million times more.

Many people who knew Cecilia would remark that she often worked until she was exhausted. But the scent of discovery was in the air, and she knew it. "In the heady atmosphere of New England," as she described it, "nothing was impossible. Once I worked for 72 hours straight without sleep."[5]

Still, it was not lost on her that she was alone—by herself in the darkened observatory, by herself as the only scientist in the world coming up with those findings. David DeVorkin, senior curator at the Smithsonian National Air and Space Museum, notes that "for any astronomer, let alone a graduate student, even at Harvard, to demonstrate that the Universe is profoundly different than previously supposed, assumed or even determined to be, would be extraordinary."[6]

Years later, that same astronomical community would offer great praise for Cecilia and her work:

> Undoubtedly the most brilliant Ph.D. thesis ever written in astronomy.[7]

> The giants—Copernicus, Newton, and Einstein—each in his turn, brought a new view of the universe. Payne's discovery of the cosmic abundance of the elements did no less.[8]

> Probably the most eminent woman astronomer of all time.[9]

But for now she felt nothing but fatigue and bewilderment. To do research, to work hard, to observe and discover—that is why she had come to America. She had fled post-Edwardian England, escaped the prevailing view that ambition was unfeminine. So far, though, it seemed as if she had simply found the freedom to be frustrated.

Worrying about money, struggling to understand the nature of the universe, facing the prospect of defying distinguished and worldly men of science—it was a lonely time. Moreover, she had not yet been able to achieve what every scientist strives for: understanding truths about the natural world. It was all enormously frustrating, for Cecilia Payne was without question a born scientist.

I

BEGINNING

Wendover and London, 1900–1919

1

It was her first encounter with astronomy. Young Cecilia was in a pram pushed by her mother, Emma, on a clear dark winter night. A brilliant meteor suddenly lit up the sky for a few seconds, bringing mother and child to a halt as they stared after it. Emma invented a little rhyme to make sure her daughter would remember what she saw:

As we were walking home that night,
we saw a shining meteorite.[1]

Cecilia's relationship with her mother was what could be described as English proper. Years later, she would characterize her mother as "prismatic and pruniferous," referring to a prim, humorless character from Charles Dickens's *Little Dorrit*.[2] Cecilia was much closer to her father, Edward. His influence on her was enormous. It was also all too brief.

Edward Payne was born in London in midsummer 1844. He followed the traditional English path of classical education, receiving degrees in Latin and Greek with honors from Oxford's Magdalen College. He was called to the British bar when he turned thirty years old. Because he was a barrister he did not have to deal directly with clients, freeing him to do research and to write. When he wasn't working, he was playing music. To unwind, he would pull his viola da gamba from its case and give lecture-recitals, playing chamber pieces by Bach, Handel, Abel.[3]

As the nineteenth century came to a close, Edward was a fifty-five-year-old successful attorney, accomplished musician, and scholar of history. His was a very full life, except for one thing. He had no wife, no children, no family to come home to at the end of the day.

British society at the time was in transition—the strict morality of Victorian England was giving way to the excesses of Edwardian England. Queen Victoria was eighty-one years old; she was spending Christmas of 1900 on the Isle of Wight, sick with rheumatism and suffering from cataracts. Her sixty-year-old son, Edward VII, would assume the throne within a year.

There were no surprises in that royal transition. King Edward had been heir apparent for decades—more than enough time for his subjects to become acquainted with his penchant for food, for fashion, for form over substance, for leisure over industry. He was a man of taste. Actually, many tastes: for caviar, truffles, oysters, lobster; for fine wine from the Continent; for hearty breakfasts and sumptuous lunches and twelve-course banquets that cost more than a maid's annual wage.

It was a time of great prosperity. From commerce to shipping, England held enormous power; London was the financial center of the world. The well-to-do engaged in conspicuous consumption and followed strict societal rules; the roles of men and women were sharply defined, and very different. Men lived in the public sphere and were expected to earn the family's living. Women were cordoned off into a private world—ornamental, subservient, more or less possessions. Because women from wealthy families were not expected to work, their success and happiness hinged on marrying well.

Although Edwardian women were gradually gaining political power—British suffragettes were taking to the streets—many women focused on attracting a husband. Skirts were cinched tight, hips scrunched into corsets. The accessory of the time was the "picture hat"—a feathered nest with a wide brim that set off a woman's face like a picture frame. Upper-class girls learned to play the piano and the harp, to speak French, and to sing, dine, drink moderately, and dance their way to marriage.

British bachelors, dressed in three-piece suits, shirts with upturned starched collars, and polka dot silk ties, would attend formal parties to look for eligible women. Girls wore their hair in braids almost to their waists. Women, on the other hand, signaled that they were ready to marry by pulling their hair up into a bun.

Because of his relatively high-profile career, Edward Payne would be on the Edwardian soirée party list. But Edward was not indulgent, like his royal namesake. He was not looking for a liaison; he wanted a wife. At fifty-five he was not interested in a society debutante. He was mature, and she had to be too. Someone at least in her early thirties. Someone, for example, like Emma Pertz.

Emma was thirty-three, a classically beautiful woman. Her father was a major in the army in Koblenz, Prussia. One can imagine Edward being drawn to her, captivated by her look—her soft eyes, her playful smile. Edward also could not have escaped noticing that Emma wore her hair pulled up. She was ready. So was he. And so it was that in 1899, Edward John Payne and Emma Leonora Helena Pertz were married.

They made a wonderful match. Edward, earnest and industrious, worked hard in his barrister's chambers in London's Stone Buildings. Emma, too, worked hard—as a painter. She did not consider herself sufficiently skilled to put her name to original works, but she was good enough to become an art copyist. She would spend three days a week creating faithful renderings of paintings at the National and Tate Galleries, mostly those of the Romantic marine landscapist J. M. W. Turner. One of her best was a copy of Turner's *Chichester Canal*, a peaceful if uninspiring earth-tone scene of a square-rigged ship at rest in still water that reflects a sunset sky. Edward often said, proudly, that "Turner improved upon Nature, and my wife improves on Turner."[4]

Emma Payne   Edward Payne

Emma was not one to waste time. Idleness was a sin for many of
her relatives. She liked to tell the story about the time her step-great-
grandmother and her five sisters—Katherine, Susan, Mary, Frances,
Leonora, and Joanna—found themselves as guests at a country house.
They were so irritated at being expected to sit still with their hands folded
in their laps that they tore all the edges off their handkerchiefs and re-
hemmed them—just for something to do.[5]

Edward was not idle, either. When it came to family life, he was behind
schedule. He may have felt a tug from the Domesday Book, the nine-
hundred-year-old tax ledger now preserved in the British National Archives. A search of the book's pages shows that for almost ten centuries,
the Payne family had been landowners in and around the Chiltern Hills.
Edward and Emma, soon after their marriage, settled into Holywell
Lodge, a rambling stone house on a quiet street in the Buckinghamshire

Holywell Lodge, Wendover

village of Wendover, forty miles northwest of London. The town name is Celtic in origin and means "white waters," a reference to the stream that runs through the village carrying chalk deposits.

Edward now had a wife and a house, but not a family; all he needed was children. A year into the marriage—May 10, 1900—Edward and Emma had their first child, a daughter. They named her Cecilia Helena Payne. A son, Humfry, was born two years later. A second daughter, Leonora, was born in 1904.

Cecilia remembered her childhood in Wendover as storybook-perfect. Everyone knew everyone else. Neighbors were a bicycle ride away. No highways, no cars, shining stars in a black sky. It was a happy home. Cecilia and her brother would watch at the window, anticipating the candlelit glow of their father's lantern swinging back and forth on a wintry night as he returned home from work in London. He would bring them small candy treasures and sit with them on the floor as they played with the blocks he had a woodworker make for them.[6]

Humfry was warm and kind-hearted; he would grow up to be a noted archaeologist. Leonora was a quiet third child; she would marry another

Four-year-old Cecilia

Londoner, Walter Ison. Together, Leonora and Walter produced an elegant book, *The Georgian Buildings of Bath*. He wrote the text, and she drew the illustrations. Cecilia, however, was different. She stood out—from everybody. Edward and Emma noticed it right away. Their firstborn was so curious—not just mildly so, but relentlessly.

One of the earliest photographs of Cecilia, taken when she was four years old, shows her standing in front of a stone table. She wears a frilly white dress and dainty little white shoes. Her hair is cut short, with bangs across her forehead, and there is a small garland of flowers placed atop her head like a tiara. The photo captures a time in which "little girls should appear innocent, virginal, unsullied in every way. Dressed in white muslin frills, they were adjured to keep clean, to keep quiet, and to keep still."[7]

The little girl in the picture is adorable. Shy, unsmiling, looking mildly uncomfortable, she is obviously posing for the camera—for keeping quiet and keeping still were not in Cecilia's blood. Her eyes were always open, watching, studying, questioning, observing. In her early years, those eyes were often looking up. Before she could read, she could point to "Charles's Wain" (that part of Ursa Major known in the United States as the Big Dipper) and Orion's Belt.

She had a restless mind and a vivid imagination. After a particularly violent summer thunderstorm, she looked outside to see the ground moving. She was captivated; it seemed as if the garden was rippling like the surface of a beautiful pond. Racing outside for a closer look, she saw that the rain had churned the soil to reveal a sea of wriggling black slugs. She cried bitterly to think that the world could create something so revolting.

Two seasons later, a rare snowstorm swept across the English countryside. To Cecilia, the pure white fluffy blanket looked as warm and inviting as the little snowsuit she was tucked into. Exasperated, Emma finally gave in to her curious daughter's demands and plunked her into a snowdrift. The unexpected icy chill brought tears, until Emma took off her shoes and rubbed her feet.[8]

In some ways, Edward was very conventional. He was a relatively well-to-do barrister; and what well-to-do London barristers did was move to the suburbs, away from the harshness and chaos of the city. But with regard to the women in his life, he was far from traditional. The prevailing attitude of the time among the upper class was that a working wife threatened her husband's manhood: "A paid job for one of his womenfolk would have cast an unbearable reflection of incompetence upon the money-getting male."[9]

But Edward apparently felt no threat. Just as he celebrated his wife's painterly talents, he reveled in and encouraged his first-born's nonstop need to know. To Cecilia, he imparted his industriousness, his passion for learning, his joy for music. "Popsy," as he called her, was only two weeks old when he began playing scales for her on a recorder. He taught her to sing—doxology hymns such as "O Sanctissima"—as he banged away on the keyboard of a miniature pipe organ. Cecilia came away from these sessions with a fine sense of pitch, to the point that, years later, she would wince when singers and string players were, to her ear, out of tune.[10]

She adored her father. At night, she would creep down the stairs from her bedroom to listen as Edward sang and played in a trio with friends. He in turn always had time for her. When she was upset, she would run to his study for comfort; he would put aside his pen, pick her up, and take her for a walk.

Edward made sure there were always one or two domestic servants in the house, freeing Emma to paint, so the family was certainly comfortable. Yet for Cecilia there was an ever-present uneasiness. As the eldest of three siblings, she was looked up to by her brother and sister, but she was also a girl. In society's eyes, Humfry was the sibling who mattered.

Decades later Cecilia still recalled in great detail the morning when Humfry's godfather drove up to Holywell Lodge in a magnificent horse-drawn carriage. His godfather asked the wide-eyed boy if he would like to go for a ride; he could be the "red-legged partridge." The boy then put on bright crimson leggings—the kind worn first in 1855 in Paris at a grand celebration staged by Napoleon III and used thereafter to signify special occasions.[11] Cecilia begged to go along, but to no avail; she could only watch as they clip-clopped away.

She ran to her father's study. Edward was deep into writing a volume of his *History of the New World Called America*. He picked up his daughter and they went for their customary walk. Coincidentally, they ended up on the carriage's return path. Edward signaled for the coachman to stop and ordered him to make room for another passenger. Cecilia described the brief journey back to the house as riding "into the seventh heaven."[12]

Cecilia with Humfry, 1904

Cecilia was lucky, for her siblings were not yet old enough to command more of Edward's attention. Edward focused intently on his eldest child as if he were making up for the lost time as an older father, and almost as if he had a premonition of the lost time to come. He had experienced heart trouble and dizziness from time to time, but the circumstances of what happened to him in December 1904, on the day after Christmas, have never been completely explained. All we know comes from an obituary in a legal journal, *The Law Times*:

> Mr. Edward John Payne, Recorder of High Wycombe, was found drowned on Monday in the canal of Wendover. . . . At 10.45 he was seen near the Perch Bridge, Halton, walking with his dog, and at 11.20 a woman named Anne Smith, wife of a pedlar, saw a hat and umbrella on the towing-path. She was then standing on the bridge, and on going to the side of the canal found a body in the water. Assistance was fetched and the body got out, Mr. Payne being quite dead.[13]

Cecilia was just four years old when her father was found in the "white waters" of Wendover. What she described as the happiest moments of her childhood were suddenly just memories. Her total inheritances from her father were his industriousness, his love of music, and his beautiful violin. Her curiosity and her determination to learn and understand remained; but to satisfy them, she was now on her own.

2

The bee orchis is a hardy orchid commonly found in England's woodlands; it grows to about a foot tall and flowers in June and July. Cecilia was eight years old when she caught sight of its distinct blossom on a breezy day in the summer of 1908. She didn't pluck it; she just gazed at it, hidden among the tall grass of the family's orchard.

She burst into the house to tell Emma the news. "My Mother could not believe me. I must be mistaken: such a flower could not be growing in our homely Buckinghamshire soil."

Emma was surprised—no doubt as much at her daughter's ability to identify a species of orchid as at the indisputable specimen growing in the grass. Emma looked at the single orchid standing out among the fruit trees and then told Deering, the gardener, to dig it up, carefully, and plant it in the garden under a little spruce tree.

Cecilia had recognized the bee orchis because her mother had once told her about this amazing plant, whose purple blossoms have a lip that looks like a bee. She was thrilled to be able to name the orchid from her mother's description. "I was dazzled by a flash of recognition," she recalled later.

"For the first time I knew the leaping of the heart, the sudden enlightenment, that were to become my passion. I think my life as a scientist began
at that point."

The new site of the transplanted bee orchis became Cecilia's secret
shrine. There, alone, she made a personal vow—she would devote herself to the study of nature, to science. With no other children her age to
play with, her imagination took over. She created "Grenson," a playmate
in her mind. A grove of trees in the garden became Grenson's Wood, a
fantasy world of two people, one real, one imagined.[1]

Emma could not speak about Edward to her children. Just the mention of his name brought a flood of tears. Cecilia, missing her father, im-
plored Emma to tell her bedtime stories. Emma complied, but not with
stories she made up. She turned instead to her extensive collection of
books. (There were no bare walls in Holywell Lodge; where there wasn't
a piece of art, there was a bookcase.) The first book she read to Cecilia,
however, was not the 1900 equivalent of *Clifford* or *Goodnight Moon*; it
was Homer's *Odyssey*. Just as Cecilia would become comfortable with nature outside, she would be at home with books inside—historical works
and literary novels in French, German, Latin, Greek, even Icelandic.[2]

It was not easy for Emma, on her own now after Edward's death. But
although she was as traditional and conventional as her paintings, she was
also resourceful. Her widow's pension was quite small, yet Emma found
ways for her children to attend plays, listen to concerts, travel around
Europe—a copyist copying the traditional trappings of the English upper
class.

By a stroke of luck, Elizabeth Edwards, a woman of Welsh descent
with a natural talent for teaching, founded a small schoolhouse across the
street from the Payne family home. "Shall I be able to read the *Encyclopedia Britannica*?" then six-year-old Cecilia had asked her.[3] Told that she
would be able to read anything, it wasn't long before she read everything:
at first simple primers, then all those books on the shelves at home. She
viewed books as foreign lands waiting to be explored, in their original
language.

Under the patient and encouraging eye of Miss Edwards, Cecilia became conversant in French, gained basic knowledge of Latin, and acquired a true command of geometry and algebra (she delighted in solving quadratic equations). She stood out physically as well. Her father's traits were emerging. She was tall for her age, and broad-shouldered. And left-handed. It made for challenges. Cecilia actually felt physical pain as she was instructed over and over to hold her pencil in her right hand. She also had difficulty distinguishing right from left. Because of her height, she was placed at the head of the line in dance class, where mistaking which foot to move first would make for constant embarrassment.[4]

Her strong left hand had become a problem. As she would do many times in the future, she began to research how to deal with it. Her great-grandfather Garth Wilkinson had been a London physician in the 1830s. His surgery practice predated anesthetics and antiseptics, and he hated it. With Henry James, Ralph Waldo Emerson, Charles Dickens, and Robert Browning as friends and contemporaries, Garth soon replaced his scalpel with a pen. He churned out numerous medical, social, and religious books and papers. In 1856, he briefly put aside those weighty topics to produce a curious little instruction manual entitled *Painting with Both Hands*. His method involved a form of mirror drawing and writing using both hands at the same time.[5] Cecilia read every word of the pamphlet and then practiced religiously, not only becoming ambidextrous but also learning how to write backward, upside-down, and upside-down-backward.[6]

In addition to teaching her students languages and mathematics, Miss Edwards schooled them in the proper English way of comporting oneself. Discipline was strict; going out for a walk was more like a military march. Correct posture during oral lessons required sitting with a backboard—a wooden board placed behind the back and held with both hands. Politeness at parties meant that "ladies and gentlemen" restrained their appetites and, no matter how ghastly the experience, always thanked their hosts.

This last rule of etiquette—thanking her host—was difficult for Cecilia. Her manners were polite, but her emotions barely controlled. Two of her classmates were the Massingham brothers: Harold, later to become a prolific writer, and Richard, an actor. Their father, Henry William Massingham, a radical London journalist whose constantly pursed lips gave him a pinched, vaguely superior air, had been forced out of his position as editor of a number of publications—*The Star*, the *Daily Chronicle*, *The Nation*. H. W., as he was known, sponsored a children's party that featured a Punch and Judy show. It was Cecilia's first exposure to the popular puppet shows in which Punch regularly beats his wife, Judy, and eventually dispatches the hangman. She was horrified.

As she left, she dutifully thanked her host, telling him how much she enjoyed the party. As she recalled later, "I can see Mr. Massingham now, his pale ascetic face set off by red hair and moustache. He looked me keenly in the eye. 'You didn't *look* as if you were enjoying yourself,' he commented drily."[7]

Each school day began with a hymn. Cecilia enjoyed the music but began to question the meaning of the lyrics. Her grammar-school mind was already becoming aware of the conflict between science and religion. And the school's religious atmosphere was stirring something else as well. "Why," she once asked Miss Edwards, "was Jesus a man and not a woman?"

"Because in his day," Miss Edwards replied, "a woman could not have done the things he had to do." The reply did not convince Cecilia and only deepened her growing sense that it was a man's world, a hurdle that would prove to be far more daunting than adapting to a right-hander's world.[8]

It was becoming clear, to both Cecilia and others around her, that she was different from her classmates. With no father in the house, the eldest child felt herself being leaned on more and more by her mother and her siblings. There were simply more things she had to do by herself than was true for her friends—a forced self-reliance that would serve her well in the future. Stories at bedtime, for example.

When she had asked for a story after being put to bed, her mother and father had always complied. But Emma now was bone weary; Cecilia was told that the routine would have to stop. "Very well," she told Emma defiantly, "I will tell stories to myself." And she did—night after night, making them up on the spot. Not just for a few weeks, but for years: she was an audience of one. And as time went by, the four-year gap in age between Cecilia and Leonora seemed to narrow. After tucking Leonora into bed, Cecilia would lie beside her, composing and reciting original fairy tales until her little sister fell asleep, imposing on herself a discipline of imaginative routine.[9]

Cecilia's ability to just "see things" would grow sharper over time. The little school across the street, led by the encouraging Welsh woman to whom Cecilia would later dedicate a published work, prepared its most famous student well in academics and manners. But its lasting effect was in furthering Cecilia's ability to observe. Once a week, the students were required to spot three brass carpet tacks that had been randomly placed somewhere in the school garden. Other exercises honed additional valuable skills. Cecilia thrived on the competitive exercises based on the complex British currency system in which students had to answer with lightning speed questions like "What is the cost of a dozen articles at eight-and-sixpence-three-farthings each?" The finest piece of scientific equipment in the school was a balance for weighing chemicals—a simple wooden beam with a pan suspended from each end—which the students were taught to use with accuracy and respect.[10]

Hands trained for measuring, eyes trained for observing, a mind trained for accuracy—one could not ask for better preparation for the study of astronomy. In time, the delicate chemical balance would be replaced by a large and complex telescope, and the little brass tacks spotted within walking distance in the garden would become atoms identified in a star hundreds of light years away.

3

"You will always be hampered by your quick power of apprehension."[1]
So wrote Miss Edwards in her farewell letter to twelve-year-old Cecilia. Emma had decided to move the family away from Wendover. Just as Emma had moved to Wendover for Edward, she felt she now had to move for Humfry. He would soon be of the age when boys traditionally entered a good English public school, and Emma wanted to be sure he was prepared.

The move had to have made Cecilia uneasy. She was an aggressive learner, but middle-class Victorians "educated boys for the world, girls for the drawing room."[2] Cecilia kept Miss Edwards's letter close. She must have been pleased that her teacher recognized that she was a quick study. But why "hampered?" She would find out.

For Cecilia, the move was like uprooting the bee orchis and transplanting it from garden soil to sidewalk cracks. She was being forced to leave hedgerows, rolling hills, bicycle paths, clear starry nights for densely packed neighborhoods, horse-bus routes, smoke-filled skies—for what she had always thought of as "a brooding shadow on the horizon."[3] For London.

The four Paynes settled into a Victorian stucco house in Bayswater, one of the city's most cosmopolitan neighborhoods. Upper-middle-class neighborhoods in London were largely homogeneous—the men leaving early each day for work in the "City," their wives managing the home, their sons preparing for boarding school, their daughters often not even going to school at all. Not so Bayswater. It had a diverse population— Greek, Arab, Brazilian, American—making it a vibrant place to live. Life at home was not quite as bleak as Cecilia had imagined it would be. The houses were packed together, to be sure, but there were at least some trees. The garden was small, but it looked out on an expansive public square. Still, as Cecilia would put it years later, "I ached for the open fields: nostalgia is not reserved for the old."[4]

She was the new kid in town; making friends was not easy for the little girl from the country. In Wendover, when she had looked to the sky, there were endless shows, from comets to constellations, playing across the dark screen overhead. Her friends then were the stars, friends that in time would reveal their secrets to her; but those friends were obscured now in urban haze. New friends would have to be of earthly variety, and Cecilia was painfully shy. The only visitors were Emma's acquaintances who came for tea.

Cecilia enrolled at St. Mary's, a strict Catholic school, where she did not find much comfort. At the end of the school day, students would assemble for daily piano recitals, after which there was nothing else to do but chat, which Cecilia found difficult. In the end-of-the-year competitions, she always carried off first prize in the essay section, but she failed at reading aloud. She was too fast a reader; her eye outpaced her tongue. In time, she would figure out a way to overcome it: she would avoid reading in front of an audience, preferring to paraphrase and improvise.[5]

Cecilia's talent for thinking, not talking, brought on other problems. All students, from every grade, were required to take an annual general

Cecilia (*right*) with Leonora and Humfry, 1912

knowledge examination. Cecilia was in the youngest grade level, yet she placed second in the school overall. The principal called everyone into assembly and berated them for "allowing" one of the youngest students to outperform them.[6] Unsurprisingly, their response was resentment, and teasing Cecilia became a school sport. She endured the teasing silently. Although she didn't know it then, it was an approach that would serve her well later when she faced scientific skepticism.

Cecilia found solace in the Bayswater home. The Wendover bookshelves, overflowing, survived the move. Most of Cecilia's ancestors, however, had been historians, and among the thousands of books there were very few on science. Poring over the family collection, she found a decades-old botany textbook on the Linnaean classification system, written in German and French. She borrowed a dictionary from the school and laboriously translated Carl Linnaeus's principles of taxonomy into English. She also uncovered Newton's *Principia*. Sir Isaac's proofs of his laws of motion and gravity were beyond her, but the assumptions were far more acceptable to her than what was taught in religion classes.[7]

In Emmanuel Swedenborg she found a prolific writer whose intellectual journey from chronicler to philosopher of chemistry and physics kindled a life-long mystical view of science. But it was in Thomas Henry Huxley, an ardent supporter of Charles Darwin, that she found a true kindred spirit. Like Cecilia, Huxley was largely self-taught. Cecilia would read and re-read his essays throughout her life, learning to develop "the spirit of a scientist" under his influence.[8]

The open inquisitive atmosphere of the little school across the street in her childhood years had been replaced now by a large, regimented Church of England institution. The stated purpose was higher learning, but religion ruled. The school had its own chapel with a service at the beginning and end of each day. Roll-call was taken in daily classes on the Bible, catechism, and Christian history. From the very start it was too much for Cecilia. She took to fainting during the services so as to be excused from chapel attendance. She once asked a London bookbinder to combine the *Apology* and the *Crito* into one volume with "Holy Bible"

Emma and Cecilia, ca. 1914

inscribed on the spine so that her teachers would think she was working on her religion studies instead of reading Plato. (The good bookbinder was indignant, and refused.)[9]

Not that she rejected the Old and New Testaments out of hand. But she saw them as examples of great literature and sources of ancient history, not as the literal word of God. Cecilia believed in numbers, not parables, so she devised a way to practice her faith: she would test the power of prayer. When it came time for finals, she divided the exams into two groups. She prayed for success in one and not the other. Sure enough, she received higher marks in the group for which she had not sought divine intervention. She knew that the result might have been preordained by her own subconscious grouping, but nonetheless, she drew the conclusion that "the only legitimate request to God is for courage."[10]

The school occupied a high and narrow townhouse. On the top floor, difficult to get to, was a chemistry lab reserved for what little teaching of science took place in the school. There, neatly arranged around all four walls, were elements and compounds in bottles of various shapes and sizes and colors. The laboratory became twelve-year-old Cecilia's chapel, where she would steal away, alone, to conduct her own worship service. "Here were the warp and woof of the world," she would write later, "a world that was later to expand into a Universe."[11]

Because the school was so focused on devotion, there simply weren't enough resources or personnel to satisfy Cecilia's yearnings. In her first year, there were no science classes at all, and the mathematics offering was not the least bit challenging. The school subscribed to the prevailing view that girls should concentrate on reading and writing and did not need to develop numerical skills. The new arrival who adored quadratic equations was placed in a class that was still wrestling with long division. Beginner algebra was a year away. The chemistry lab was officially off limits.

Cecilia needed another teacher, another mentor to channel her consuming need to learn. Not a moment too soon, Miss Daglish came on the scene. Young, witty, and full of energy, Dorothy Daglish was hired by the lower school to teach science—indeed, all of the sciences. The first year she taught botany, and she found in Cecilia a budding botanist who needed direction.

Cecilia had been designing her own do-it-yourself science projects. For the past year she had been putting the family's holiday walks to good use by collecting plants of all kinds and bringing them home to dry. She created a herbarium, identifying each species per the classification system of Carl Linnaeus's two-hundred-year-old textbook.

Dorothy immediately recognized the young girl's genuine passion for science, and she wasted no time. The Linnaean system? Dorothy believed it to be largely obsolete. And the homemade herbarium? She pronounced it "dried hay."[12]

It was tough love, but it was love. Dorothy began to instill discipline in her new student by encouraging Cecilia to replace the dried hay with drawings. With Emma organizing and leading the way, the family would go on holiday expeditions, all sitting in a row and sketching the view. Unlike her mother, Cecilia was not a good illustrator. She was not nearly as accomplished as her brother Humfry, who would become a fine painter in addition to an archaeologist, or her sister Leonora, who would become an architect. But Cecilia was game, and eventually she compiled a portfolio of drawings that helped her develop a detailed knowledge of systematic botany and furthered her ability to recognize patterns.[13]

As she moved from teacher to mentor and finally to friend, Dorothy freed them both from the confines of the classroom. She taught Cecilia about the compounds in the chemistry lab's bottles, introduced her to books on physics, and took her to museums. One Christmas, she told Cecilia she would lend her a book about astronomy but then decided to give it to her as a gift instead. She could not have known what a prophetic gesture it would be.[14]

Cecilia valued books and all things scientific but had little interest in material wealth, which she had come to associate with her dreadful great-aunt Florence. Florence Wilkinson, her maternal grandmother's sister, had married into great wealth. Her husband, Benjamin St. John

Attwood-Mathews, was one of the founders of England's Alpine Club, the world's first mountaineering club.

Florence reigned over Llanvihangel Court, her husband's ancient family home in Monmouthshire. Queen Elizabeth had once slept there, and supposedly there was a bloody footprint on the stairs left over from a long-ago duel.

Cecilia paid one, and only one, visit to her great aunt. She would have approached the estate along its splendid avenues—planted with chestnuts said to have come from captured Spanish Armada ships—and then walked carefully through the paneled entrance hall past the two over-sized jars containing wine from the same ships.[15] On a throne-like chair waited Florence, clad in black satin, as stout as the obese little pug sitting on her lap.

It was Florence's well-known attempt to use her wealth to render her sister's descendants subservient that gave Cecilia "an abiding horror of money."[16] During their conversation that morning, Florence gave Cecilia advice on whom to marry: specifically, an elderly bishop who was a distant cousin of Florence's late husband. Cecilia, of course, had never met the gentleman. "I told her stoutly," Cecilia recalled later, "that I was perfectly capable of earning my own living."[17]

Cecilia was not a model post-Edwardian girl. She was not going to be told whom to marry, not going to be told she couldn't study science. She was increasingly independent-minded. And her interests expanded beyond science. When she wasn't working with Dorothy, she could often be found at the Old Vic.

The Royal Victoria Theater, built in 1818, was and still is a sturdy structure just southeast of Waterloo Station. Renowned for bringing high art to the masses, the theater began a series of Shakespeare performances just as Cecilia turned fourteen. The Paynes "haunted the Old Vic" week after week, paying pennies to see *Othello*, *Richard II*, *The Merchant of Venice*, *The Tempest*. The family also attended operas and classical music concerts at Albert Hall. Emma kept Edward's memory alive by constantly talking about the composers he loved, particularly Handel. When the chorus stood to sing at a performance of Handel's *Messiah*, Cecilia burst into tears. Art was just as emotional for her as science.[18]

For a trio of creative siblings, it is not surprising that the Payne children made the leap from attending to performing. There was little sibling rivalry among them—the first-born was energetic and in charge. To put the local church to what she considered to be better use, Cecilia volunteered to give Shakespeare performances to children from nearby poor neighborhoods, starring herself and her brother and sister. She chose scenes that required just three actors. In the mode of their industrious forebears, they made their own props out of muslin, tinsel, and cardboard. She described the theater, the opera, and the ballet as the "lifeblood" of her childhood. (Years later she would write, "I pass lonely hours by repeating to myself scene after scene of *Hamlet*, which I committed to memory at that time.") It was another obstacle overcome: acting out a prepared script, she had found a way to speak in public.[19]

With the amateur three-person Shakespeare troupe a success, Cecilia used the same audience for another purpose. She had been told that to fulfill her dream of becoming a scientist she would be required to be a teacher. She quickly volunteered yet again—to teach these same children Sunday School. But she did it her way, with more science than scripture. She kept, and treasured, a cartoon that Humfry drew at the time, of her addressing a group of wide-eyed, utterly terrified young boys. The caption read, "And now that you know all about the photosynthetical effects of convection on carbon dioxide, we will go on . . . ."[20]

Word of Cecilia's Sunday "sermons" did not sit well at St. Mary's. She was becoming a problem in the eyes of the school's staff. And it was all happening just at the moment when her situation changed for the worse in other ways. As World War I was breaking out, Dorothy Daglish fell ill and had to stop teaching. The war effort quickly consumed men and women schooled in physics and chemistry. Science teachers were hard to find, which left Cecilia to learn on her own once again. Most of the staff were language teachers, and they were determined that Cecilia would be trained as a classical scholar. No one would describe her as indifferent to the classics; she just wanted to learn botany, physics, and chemistry as well.

When she turned fifteen, she won an academic contest; the prize was a book of her choice. The staff expected her to choose Shakespeare, or perhaps Milton. She said she wanted a textbook on fungi. The administrators were aghast, but she was not to be denied. She eventually got it— elegantly bound in leather, befitting Shakespeare, or perhaps Milton.[21]

If she had to learn by herself, so be it. She avidly read the up-to-date botany textbooks that Dorothy Daglish had introduced her to. (She recounted later that she learned about sex not from ever proper Emma, but rather by working it out herself as she studied the pollination of tropical cycads.)[22]

Toward the end of the year came the college preparatory exams, with the standard sections on mathematics, literature, and languages. Cecilia asked to take the test on botany as well. The school administration knew that botany was not formally taught. It was a nationwide test, and they probably did not want one of their students to embarrass the school's good name; they said no. Cecilia prevailed. When the botany test results were posted, Cecilia's name was at the top of the list.[23]

Still she wanted more. She knew that to become a scientist, she would need to be fluent in German and proficient in advanced mathematics. It was a lonely battle—no other girl expressed any such desire. She was so different from her classmates, who were from aristocratic families and were being groomed to take their place in that world. Several would become successful actresses. Others would go on to a finishing school, where the curriculum focused on how to dance, how to enunciate clearly, how to comport oneself at banquets, and, most importantly, "how to enter or retire from a room with a degree of elegance and assurance."[24]

Cecilia began studying calculus and coordinate geometry by herself. She also tried her hand at writing poetry, such as this rather pointed verse about one of her teachers:

Out on you, fond instructor, perverter of Nature's laws,
Explaining cause by effect, confounding effect with cause!
I could say a thousand things about you, but I will desist,
For you are a charming woman, but you are not a scientist.[25]

The school's teachers and administrators did not know what to do with her. She was not compliant, like most of the other girls. And Cecilia did not help her case. Impatient with teachers who could not keep up with her, she would remember later what a "hard time" she gave them. "I suppose teachers of science were difficult to find," she wrote, "and I was not slow to realize how easily they could be confounded."[26]

The school had a choice: accommodate her need to learn, or ask her to leave. At first, they tried accommodating. The school administrators did not want to hire someone to teach just one student, so an English teacher volunteered to tutor Cecilia in German. Another, a humorless algebra teacher, was asked to tutor Cecilia in advanced mathematics. But the woman interpreted Cecilia's passion for math as passion for *her*. When Cecilia made clear that her strong feelings were just for the math, the woman sneered, "You will never become a scholar!"[27]

That did it. Just days before the end of the school year, days before her seventeenth birthday, Cecilia was summoned to the office of the principal, "a saintly Churchwoman," as Cecilia later described her. She told Cecilia, "You are prostituting your gifts." The words were meant to shock, and they did. She was told she had to leave the school.[28]

The school had finally given up on her. Cecilia's goal was to go to Cambridge, but her family could not afford the expenses. She wanted desperately to be a scientist, but she had only one year of preparatory school left, and the only subjects in which she had received formal training were Latin and Greek. Her one hope had been to learn advanced math, and that had failed. Not to mention that the world was at war. Her dream of becoming a scientist seemed to be just that.

4

It was not supposed to be like this. Humfry was supposed to be the center of attention, not Cecilia. Boys were to be educated, girls refined—and Emma had uprooted the family from Wendover to move to London so that her son could get into the right school. But now the unexpected had happened—Cecilia had been expelled. Another school for her had to be found, and quickly.

Cecilia had little time for chit-chat, little patience for finery or socializing, and no interest whatsoever in material wealth other than books. At seventeen, she was very much her own person. She was approaching five foot ten now—taller than all the other girls, taller than most boys. Gone were the little-girl bangs; now she wore her hair pulled straight back. She looked exactly like what she was: a tall-for-her-age attractive young woman . . . on a mission.

It was a mission facing strong headwinds. A schoolgirl's thirst for science was not something that English society was comfortable with. One headwind came from the direction of the clergy. Henry Liddon, a theologian with great influence over the Church of England, declared the idea of higher education for women to be a "development which runs counter to the wisdom and experience of all the centuries of Christendom."[1]

Another came from the medical profession, which made the case that higher education had detrimental effects on the "less robust" gender. "Before sanctioning the proposal to subject woman to a system of mental training which has been framed and adapted for men," wrote the psychiatrist Henry Maudsley, "it is needful to consider whether this can be done without serious injury to her health and strength."[2]

And from the academic community came Herbert Spencer, the British social philosopher who coined the phrase "survival of the fittest." He argued that a woman's primary biological role was to be a mother, and that "women who claimed educational opportunities equal to men would 'overtax' their brains and this would diminish their ability to bear children."[3]

Fortunately for Cecilia, Frances Gray paid no attention to such pronouncements. Chosen out of sixty-seven candidates, Frances Ralph Gray was the founding "high mistress" of St. Paul's School for Girls, in the Brook Green neighborhood of London. She had enrolled forty years earlier at Newnham College at Cambridge, passing examinations with distinction in English history and literature, Latin and Greek, and political economics. Like Cecilia, she adored music; unlike Cecilia, she had had great difficulty learning mathematics. Also unlike Cecilia, she was small in stature. She was, however, commanding in presence. Students reported that being sent to see Miss Gray "was their greatest fear."[4]

Frances perceived that there was something different about the seventeen-year-old girl in her office who so wanted to be admitted. Yes, Cecilia had been told to leave her current school, but not because she was disruptive or a problem learner—quite the opposite. She was a serious student who loved music and science, whose goal was to go to Cambridge. She had responded to Elizabeth Edwards and to Dorothy Daglish. If St. Paul's had similar teachers who could recognize Cecilia's love of learning and would take the time to nurture her, surely she would be a good fit.

It is not known whether Emma prevailed on the principal of Cecilia's former school to put in a good word for her, or whether the "saintly Churchwoman" felt compelled to do right by the young girl she had dismissed with just one year of school remaining. Whatever the reason, Cecilia needed help, and she got it from this unlikely source.

Years later, in a touching letter recommending Cecilia for a fellowship at Harvard, Frances wrote: "It is not my practice to admit girls who have reached the age at which Cecilia Payne was admitted [age seventeen], but I was requested to make an exception in her case by the headmistress of the School she had previously attended, who assured me that she was a girl of very unusual promise."[5]

Unbeknownst to Cecilia, St. Paul's needed her as much as she needed the school. Founded just over a decade earlier by the Worshipful Company of Mercers, it prided itself on consistently outperforming other schools. The "Paulinas" were not viewed as, or trained to be, socialites; this was a serious school. The social snobbery of other private schools had no place here.

Cecilia described her move to St. Paul's as stepping from medieval times into the modern day. Instead of chapels there were laboratories—in biology, chemistry, physics—and teachers who were specialists. Here she was not just "allowed" to study science; she was encouraged. She only attended the school for one year. But from the moment she approached the Queen Anne–style pink brick building and walked up the stone steps and through the marble and oak arched front door, she was home. "I shall never be lonely again," she said to herself. "Now I can think about science!"[6]

She may have thought about science above everything else, but she was still a teenage girl, and boys occupied a part—even if only a small part—of her mind. Her social life barely existed, however, for several reasons. For one, the social code of the time made it difficult. If her brother

St. Paul's Girls' School, London

Humfry brought a friend home, for example, his sister was expected to make herself scarce. The one time she did try conversing with one of Humfry's friends was a disaster. The boy later remarked to Humfry, "Fancy! A girl who *reads Plato for pleasure*!"[7]

If shyness were not enough (she found coming-out parties "a concentrated agony"), her clothes made things even worse. She mostly wore hand-me-downs from the daughter of one of Emma's wealthy friends. "I still remember my horror," Cecilia wrote later, "when I learned that one of my dancing partners knew her, and thought with crimson shame that he probably recognized the dress I was wearing."[8]

What the new girl at St. Paul's did do was to plunge, expertly and headlong, into public speaking. It was the blossoming of Cecilia Payne. The budding scientist in her, and the professor to come, gave prepared talks to an audience of her schoolmates. The student newspaper reported that on December 17, 1918, "Cecilia Payne gave us a most excellent lecture on 'Sound and its Transmission.' We are very grateful for the trouble Cecilia took in preparing the lecture, and explaining everything so clearly to us; and also for arranging some very beautiful illustrative experiments."[9]

The month before, noted the newspaper, she had "read a very interesting paper on 'Aviation.' Cecilia helped us to understand something of the principles of Flying, and to make our observation of the passing aeroplane rather more intelligent than before. The questions with which she was bombarded after the paper must have shown Cecilia the interest she had aroused in us, and our appreciation of the work she had put into her paper."[10]

The topic on March 21 was classic Cecilia. On that day, "a discussion was held. The motion was: 'That a knowledge of history and literature is of more use to the average citizen than a knowledge of natural science.' The proposer was Evelyn Parker, and the opposer was Cecilia Payne."[11] Cecilia won the debate by a ten-vote margin.

Nor did she feel constrained to speak just on scientific subjects. Thanks to Emma's caretaking of the family's book collection, Cecilia was extremely well read. Before the Literary Society's members, "Cecilia Payne read an excellent paper on the 'Epic.' The discussion which followed was rather halting, owing to the Society's never-failing tendency to run after the proverbial red herring. Cecilia, however, ably defended her theory that 'epic' is a name which should be given only to truly great work, and does not merely imply, as many present were inclined to believe, 'a long poem.'"[12]

Frances Gray believed that higher education should embrace more than just laboratories and classrooms. As she wrote in a 1925 letter to a school trustee: "While we shall certainly send many of our girls to University and equip them in various ways for earning a living, [parents] need not be afraid to trust us to give them the refinements of life."[13] She introduced St. Paul's students to a range of nonacademic interests, from sports to bookbinding, persuaded the board of governors to build a swimming pool, and attracted high-caliber teachers with good pay, a residence, and a pension.

She made good on her professed love of music when she hired Gustav Holst. Holst was a relatively unknown trombone player when he accepted the job of director of music at St. Paul's. Like Cecilia, he was shy and reserved, and he disdained fame. And like Cecilia, he was practiced in overcoming obstacles: neuritis in his right arm had forced him to stop playing the trombone and the piano, so he had turned to composing.

Frances encouraged him; in fact, she worked with him, supplying the text for both a light-hearted masked dance in 1909 and a more ambitious orchestral work three years later. She had an entire music wing built in 1913, including a large soundproof room where Gustav composed on Sundays, when the school was locked up, in silence and solitude. It was in this room that he wrote his most famous work, the orchestral suite *The Planets*. Cecilia was among a group of students who heard it performed shortly after it was composed.

Holst was also a great teacher. For three decades—from 1905 until his death in 1934—"Gussie," as he was known, would cast his musical spell over his students. The contemporary composer Ralph Vaughan Williams described Holst's long tenure at St. Paul's: "He did away with the childish sentimentality that schoolgirls were supposed to appreciate and substituted Bach and Vittoria; a splendid background for immature minds."[14]

Holst discerned Cecilia's love of music. He asked her to play her violin for him, made her a member of the school's orchestra, and taught her how to conduct. He encouraged her to become a musician but did not prevail. Cecilia instinctively felt that a career in music would control her; as a scientist, she would be in control.[15]

If Frances had had any reservations about admitting Cecilia, the avid science student soon dispelled them. Frances would later write:

I soon saw for myself . . . that Cecilia Payne had originality and ability far in advance of the originality and ability of the ordinary clever school-girl. I was very much struck, also, by her power of presenting a difficult subject in a lucid and attractive manner. I heard her give to the Science Club of the School a remarkably clear and interesting lecture on Aviation. It would not have been unworthy of an experienced teacher of Physics.[16]

If Cecilia was able to lecture like an experienced teacher of physics, perhaps it was because she was being instructed by a skillful teacher: Ivy Pendlebury. Just as Dorothy Daglish four years earlier had been more than a teacher, "Miss Pendlebury" was more than a classroom instructor. She was a real scientist. She had been recruited to join St. Paul's in 1911 from the London School of Medicine for Women, where she had been teaching physics.[17] In a school meeting dated the year Cecilia arrived, Frances Gray reported that physics was generally neglected in girls' schools, but "I am glad to say that physics is a subject of serious study in this school."[18]

Ivy didn't teach Cecilia in the traditional lecture method. She took her by the intellectual hand and led her through the labyrinths of physical science. Instead of giving Cecilia the answers, she allowed her to discover truths for herself—to analyze and conclude based on observations.

There were still many obstacles to overcome. Cecilia had a self-taught head start in botany, but she was far behind in chemistry and physics. Cram, cram, cram: mechanics and rigid dynamics, the Newtonian equations of motion, electricity and magnetism, thermodynamics, and basic astronomy. It was a wild wonderful year, so different from the religious emphasis of the past, so stuffed with science. The more Ivy threw at her, the more Cecilia wanted. "[Miss Pendlebury] told me that she had never had a pupil with my power of sustained application," wrote Cecilia in her memoir. "It was in fact the releasing of years of pent-up, unsatisfied desire. By the end of my schooldays, physics was replacing botany in my affections."[19]

To reach her goal of going to Cambridge, Cecilia would need to pass the difficult and competitive Cambridge admittance exam. Known as the "Little-Go," it was tilted more toward the classics than science. For Cecilia it made no difference. Indeed, one can imagine that she smiled when she saw the classics section—a choice between translating from Greek either the Gospel of St. Matthew or the *Apology* of Plato. She had studied the *Apology* and knew it almost by heart. She chose Plato.

"But you're not a Jew!" exclaimed the examiner. Cecilia was speechless. She was not worldly enough to know that the academic world distinguished between Jew and Gentile. Plato's *Apology* was offered as the option for non-Christians. It was a strict policy; no exceptions. She would be tested on St. Matthew. Ironically, all that religious training finally had a payoff: she remembered enough of the English version of the Gospel to be able to produce a passable translation.[20]

The last obstacle, of course, was money. There was only one scholarship, the Mary Ewart Scholarship for Natural Sciences, that would provide enough funds to pay her expenses.[21] And there were surely many competitors. Cecilia had spent the past year preparing for it. On the morning of the test Ivy told her nervous student with confidence, "You'll get that scholarship." The day before, she had tutored Cecilia in the theory and construction of an air thermometer. "It had been foreseen," wrote Cecilia later. "I was not in the least surprised when the examiners made the air thermometer the basis of my laboratory test."[22]

The scholarship was hers. Cambridge was hers. How did Ivy know?

II

PREPARING

Cambridge, 1919–1923

5

The train rolled steadily north from London. In the fall of 1919, Cecilia's view out the window would have looked much like it does today: the wind roiling the grass, making the various shades of green look like the surface of the sea; the occasional plot of rapeseed—England's "fields of gold"—breaking up the wash of green; the clouds, *always* clouds, lying low on the horizon. The station signs whipped past—Royston, Baldock, Letchworth, Stevenage, Finsbury Park—as the train drew closer to its destination: Cambridge.

Cecilia never thought of herself as "like the other girls." She was barely aware of, much less motivated by, vanity or fashion or social trends. What drove her was her restless mind. As the crowd of students chattered around her, she was most likely looking out the window and wondering about relativity. Was she really moving over the still earth, or was she the motionless one while the green and gold landscape slid by?

Her mind had begun to explore beyond the classroom. During her year at St. Paul's, she had learned the fundamental principle that all motion is relative. But it was while walking down a London street toward the end of that year that she had stopped suddenly, as she was wont to do. "I asked myself: 'relative to *what*?' The solid ground failed beneath my feet. With the familiar leaping of the heart I had my first sense of the Cosmos."[1]

She had tried to describe her thoughts to Ivy Pendlebury, but she couldn't find the right words. As would often happen, her mind was racing past her ability to express herself. Ivy was calm. Just as she somehow knew Cecilia would win the scholarship, she also understood what Cecilia was trying to get across. "You will find relativity very interesting," was all she said.[2]

It would indeed be an experiment to confirm Einstein's theory of relativity that would set Cecilia on the path to discovery. But in time. For now, she was just another nineteen-year-old "fresher" caught up in the pandemonium of Cambridge station. Those with enough cash clambered aboard a "growler," a four-wheeled hansom cab. Cecilia was not in that crowd; she waited, suitcases in hand, in a long line for a tram. Pulled by a single horse, it moved slowly away from the station along Hills Road. After a short journey, she changed to another tram bound for Lensfield Road and Trumpington Street.[3] It was only a mile and a half to Newnham, the women's college at the University of Cambridge, but it seemed longer.

There was so much energy around her. There cannot have been many other moments in history in which the conditions for scientific fervor were as intense as those in Cambridge when Cecilia arrived. Electricity and radioactivity were in the air, literally and figuratively. Pent-up scientific discoveries were about to burst forth.

During the Great War, men of science had been taken away from their labs and assigned to war rooms, intelligence offices, even trenches. They were diverted from research to military strategizing. For years they had not been able to theorize, experiment, or observe. So when they returned to Cambridge—the lab, the classroom, the library, the observatory— there was a slingshot effect. A devastating world war had just ended, economies were improving, and there was a heady sense of limitless possibilities.

Cecilia before entering Cambridge, 1919

As a woman entering Cambridge in 1919, Cecilia's timing was near perfect. Not only were scientific breakthroughs set to happen, but politics and mores were changing as well. Months earlier, in December 1918, she was among a sea of onlookers in London as US president Woodrow Wilson, in a horse-drawn carriage with his wife, Edith, and Prince Arthur, the duke of Connaught, had paraded past. The war over, Wilson was promoting the League of Nations, his grand vision of a cooperative community designed to make certain that a world war would never happen again.

"I had stood in a London street and heard the delirious acclamation as Woodrow Wilson rode by, waving to crowds and smiling a fixed, sardonic smile," Cecilia remembered later. "The League of Nations was to save the world, and he was its prophet. Humanity was saved, and now we could turn to other things."[4]

Though Cecilia was too young to vote, and, as a woman, would not have been able to even if she had been old enough, she was still caught up in the heady postwar scene. Because of the war effort, women not only worked right alongside men but often replaced them. Gone were conventional restraints. What was at first practical had now become fashionable: women wore their hair short; they wore their skirts and shorts at knee-length; they smoked the same cigarettes as men. "An easy camaraderie, born of work side by side, [had] developed between the sexes. The language of the trenches became the common language of the young."[5]

She probably did not know it at the time, but Cecilia had left the "brooding shadow" of London for good. She would visit the city from time to time—to see her mother, to study in the city's great libraries and museums, and to attend lectures in various professional societies. But it would never be home again. Like all 1919 freshers, Cecilia was no longer a child living at home. She had an ear for the "language of the trenches" and chain smoked. She was out of the house, away from Emma and the family, completely independent now.

She finally had what Virginia Woolf would soon so famously describe as "a room of one's own." Cecilia's was in Newnham's Clough Hall. With the lofty ornamental ceiling of its entrance hall ringed by glossy white balconies, Clough was the largest and brightest of the three Newnham residential halls. Her room was equally bright and cheerful, with two windows looking out on the gardens. The sunlight of a Cambridge autumn flooded the room with color: the golden maple trees in the distance, the silver birches closer in, the scarlet Virginia creeper climbing the wall next door, the polished mahogany chestnuts scattered on the sidewalks and garden paths.

"But even more precious than anything outside the room was the room itself," wrote a young Newnhamite at the time. "To have a room of one's own—that was the supreme pleasure, the unspoilable joy of being at Newnham."[6] The furniture was comfortable and functional, not fussy: a bed, a table, two upright chairs, an armchair, and of course a writing desk. There was a washstand in the corner behind a screen, and a large oak chest for storing clothes. Modern conveniences had not arrived, however; neither coat hangers nor the electric iron existed yet.[7]

Being on one's own meant being responsible, and throughout the college's history many freshers arrived with little to no experience in looking after themselves. "I boiled some water for my hot water bottle last night," wrote a young Newnham woman in 1889. "I do not understand the management of a kettle and the wretched thing boiled over and made an awful smell."[8]

Winter comes early to Cambridge, and the one task Cecilia and every other Newnham fresher had to learn to do, and fast, was to start and tend a fire. There was no central heating; each room prominently featured an open coal-burning fireplace. It made for a cozy setting, once it finally got going. But starting a coal fire is notoriously difficult, as new Newnhamites soon discovered. Women students in the early years were known to try coaxing a fire with paper and sifted sugar. (When exposed to flame, sugar immediately catches fire.) It was not the safest routine. Teams of women in fire helmets would shoulder fire hoses and haul them to dormitory rooftops in mandatory, and frequent, fire drills.

By the time Cecilia arrived, fire-lighters—compact bundles of wood shavings and sticks—had replaced sifted sugar. They worked well, prompting one Newnham woman at the time to shun weighty topics such as radioactivity and electromagnetism and instead propose a "Clough Freshers' debate: 'That the greatest invention science has given us is the fire-lighter.'"[9]

Each morning in the corridors of Clough there was a familiar sight: "A stalwart ex-naval rating named Mr. Bowen . . . pushing along his trolley laden with coal for our scuttles."[10] The problem was the coal itself; it didn't burn for long, and the daily allowance was not sufficient to keep a fire going for more than a couple of hours.

Some freshers solved the problem by pooling their coal; as if at a Cambridge campfire, they wrapped themselves in rugs and joined together "at one little fire in somebody's room." Others summoned the fire within, vigorously dancing Scottish reels and the Highland fling to stay warm. And then there were those who simply threw up their hands and "worked in unheated rooms in sleeping bags and all our outdoor warm clothes."[11]

From day one Newnham presented an odd contrast: the quality of the learning was world class; the quality of the food was not. In fact, it was terrible. One of Cecilia's classmates declared that the college's food "was very poor, consisting largely of semolina in various forms, and strange unidentifiable vegetarian dishes, of which the most revolting was the one known to us as a 'mess of lentil pottage.'" The staff would gamely set an example, "encouraging us to force it down our reluctant throats."[12]

The Clough dining hall, where the reluctant throats took all their meals, was the very picture of Old World formality. The staff, and a rotating group of invited students, sat at a long table on a platform raised two steps above the main floor. The rest of the students were seated in high-backed chairs at long narrow tables in three rows. The entire assemblage ate under the stern watch of portraits of principals past. It could be a scary experience, with "the noise of so many knives and forks on the crimson-colored china quite intimidating." The food may not have been all that appetizing, but there was a lot of it. The three-course meals in the dining hall were followed by late-night "Cambridge crumpets, oozing with butter."[13]

A certain Mr. Smith was in charge of checking in latecomers. His office was in the main gateway—Cecilia would have passed by him every day—a central location allowing him to observe everything. A sour little man, he was called Ignatius, for apparently no other reason than that of it being a catchier name than "Mr. Smith." As one student recalled, "it was he, they say, who was asked by a visitor what the young ladies did all day and replied gloomily, 'They eats and eats, with intervals for meals.'"[14]

Life was hardly harsh, however. The only real job required of Cecilia was to make her bed—on Sundays. For the rest of the week, an army of maids, wearing print dresses and large white aprons, would sweep into the rooms first thing in the morning to wash out the china, ward off the cold with a can of hot water, and then "kneel at the fireplace with a clatter of fire-irons to clear out the ashes and re-lay our fires."[15]

Later in the day, the maids would change into black dresses with small fancy aprons and frilled mob-caps. A student who complained about the winter weather to one long-time maid reported her reply: "If you'd done as much work as I have today, you wouldn't be cold, miss."[16] The fee for maid service (and for four meals a day) was £33 per term (almost £1,700 today).

The maids must have had their hands full with Cecilia. It wasn't that she was spoiled or felt entitled; it was simply that focusing on science left little room for attention to anything else. As another fresher and Cecilia's dear friend, Betty Leaf, wrote in a profile of Cecilia for *Thersites*, Newnham's literary magazine, Cecilia's "interest in material tidiness is chiefly theoretical, and she has a feeling of slight surprise when she loses an important letter which she has 'been carefully keeping on the floor.'"[17]

The freshers at Cambridge were like freshmen at any university. They were comrades in arms at a time of life when true friendships are made. So it was with Betty and Cecilia. Their rooms were close by; although they were very different from each other, their bond was quick, and real.

One indication of the differences between them is evident in the contrast between their biographical index cards. These index cards are now in the Newnham Archives, one for every student in the college's history. The cards are handwritten records of thousands of students, from the day they started at the university and continuing throughout their lives. While most of the biographical cards are relatively sparse, Cecilia's is crammed with information, the chronological entries written in different hands over the years. The "scholarship" box alone is completely filled—listing her as a "Mary Ewart 1st year scholar" in 1919, an "Arthur Hugh Clough Scholar" in 1922, a "Bathurst Scholar" in 1923.

Betty's biographical card is as pristine and empty as Cecilia's is messy and jammed. She was the daughter of Cecil H. Leaf, a prominent physician and graduate of Cambridge's Trinity College who had written the definitive text on lymphatic glands. Betty had graduated from affluent Kensington High School and was concentrating in English.

Betty was more than just a close friend; she was more like family. Cecilia did not have much contact with home. Having fended for herself—studying hard to be admitted, scrambling for ways to pay the tuition—she was a family outlier. Most middle-class girls her age were being groomed for marriage and did not go to college, much less to Cambridge. Emma still had her hands full, with two children at home. Humfry and Leonora didn't write.

Betty was captivated by Cecilia; she quickly saw how different her friend was from the other girls in the dorm. Like a reporter, Betty described Cecilia's quirkiness in the profile she wrote for *Thersites*, noting that Cecilia would contort herself into different positions that matched whatever her nonstop curiosity happened to be focused on. "She is always thinking of *something*," Betty wrote, "and when safely lying on her back on the floor (she despises armchairs), she will talk of all things under the sun, from ethics to a new theory of making cocoa."

Roll

Payne   Cecilia Helena   (calls herself Payne-Gaposchkin)
GAPOSCHKIN

DATE OF BIRTH	DATES OF RESIDENCE	HALL
10 May 1900	1919 – 23	Clough

SCHOOL OR UNIVERSITY

St Mary's College, Paddington
St Paul's School for Girls

SCHOLARSHIPS

Mary Ewart 1st year Scholar 1919
Arthur Hugh Clough Scholar 1922
Bathurst Student 1923.

EXAMINATIONS PASSED IN COLLEGE

Natural Science Tripos Part I Class I 1921
     "   "   "   II "   II 1923

DEGREES

B.A. (Camb.)
Pt. D. (Radcliffe)
(M.A. (Cambridge) 1952
Sc.D (Cambridge) 1952
D.Sc (Wilton)
D.Sc (Smith)

ADDRESS ON ENTRY

70 Lansdown Road, London W.

MARRIED

M. Sergei Gaposchkin March 5 1934
                        (at New York)

APPOINTMENTS   See Register & File

1923 – 4 Graduate Student, B.C.; Research
          Student at Harvard Univ. Observatory
1924 – 5 Rose Sidgwick Fellow & Pickering
          Fellow at Harvard Univ. Observatory
1925 – 6 Pickering Fellow "   "   "
1926 –   Research Fellow "   "
1935   Cannon Prize of American
          Astronomical
1936   Memb. American Philosophical
          Society
1939 – 58 Philips Astronomer, Harvard Univ.
1949 –   Distinguished Achievement Reading, The
          Graduate Chapter, Radcliffe Alumnae Assoc.
1957. Annual Achievement Award of American
      Assoc. of University Women. (Boston)

SUBSEQUENT EXAMINATIONS

7 December 1979

Died 1980 ☒

Cecilia's biographical card, Newnham College

When Cecilia, working long hours in her drafty dorm room, would come down with a cold, "she sits up in bed with her fiddle, and declares she has been spending the morning at the opera."

Now and then Betty would find Cecilia sitting cross-legged on the floor of her room, reverting to the scenery-making pastimes of her youth: "When even books and music fail, she takes refuge in the construction of mechanical toys, and a confusion of cardboard and cotton will soon turn into a figure of Icarus, who waves expressive legs, arms, and wings."[18]

Many Newnham women had lived through lonely schooldays. Now, together in a male-dominated college, they took comfort in each other's company. Fresher friendships would endure long past graduation. So it was with great affection that Betty recorded that Cecilia "refuses to go to bed until any stray thoughts provoked by some chance discussion have been properly pigeon-holed."[19]

There were indeed stray thoughts to be pigeon-holed, for Cecilia was unquestionably one who despised "a day of small things," the striking phrase Florence Nightingale had used in 1852 to describe the life of a typical upper-class English woman.[20] Yes, the maids kept her room orderly and the staff cooked her meals. But when it came to the serious task of learning, she was on her own. As another young woman described it, "We were no longer schoolgirls but adults, and we were proud of our independence. . . . If a student proved unequal to the strain involved . . . she was invited to retire as unsuited to University life." Little support was provided: "If we could not swim we must drown."[21]

For Cecilia, as intelligent and accomplished as she was, Cambridge could be intimidating. But her days were full. And after all the pigeon-holes were filled, she would climb into bed under thick covers, no doubt agreeing with another student just a year ahead who recalled: "One of my great joys . . . was stirring up the logs last thing at night, . . . turning out the main light and going to sleep with the flicker of the firelight."[22]

6

Cecilia had dreamed of going to Cambridge for about as long as she could remember. It was the promised land for a young mind inclined to science—its laboratories legendary, its scientists storied. She had spent countless hours preparing herself to get there. And, in many ways, the university had been preparing itself for her.

Roughly eight hundred years ago, a group of enterprising people built a bridge over the River Cam, fifty miles north of London. The convenience of a bridge attracted more people, and then more people. In a relatively short time, there was a critical mass to form a town. They called it, appropriately enough, Cambridge.

At first, most of the major institutions were religious, the clergymen a peaceful, orderly lot who manned the abundance of churches along the river. But beginning around 1200, two other groups discovered Cambridge: traders seeking access to the sea, and scholars from Oxford fleeing the hostility of the townspeople. It would prove to be a volatile mix. The traders brought great wealth to the town; the masters and students brought, well, rowdiness. "From the start there was friction between the town and the students," notes the University of Cambridge's official website. Students "often caused disturbances; citizens of the town, on the other hand, were known to overcharge for rooms and food."[1]

Despite the town-gown friction—to the point where resentment boiled over into actual attacks on university property—the town gradually became more of an academic center than a trading post. Even the town's physical appearance changed. Trinity College was formed in 1546 when the university took over a cluster of small houses. Large numbers of wealthy "lay students" began flocking to town, bringing with them their servants, tailors, fencing masters, riding coaches—even gardeners who tended their tennis courts.

In contrast to Oxford, where the focus was on liberal arts, science was emphasized at Cambridge. While public order kept declining, course offerings kept increasing. More and more, mathematics came to the fore, culminating in the seventeenth-century brilliance of Sir Isaac Newton. Soon, private donors began to endow professorships in chemistry, anatomy, botany, geology, astronomy, and experimental philosophy.

But there was one element of university life missing from the Cambridge campus: women. "Nasty forward minxes"—that was how Adam Sedgwick, Cambridge professor of geology, in 1865 referred to women who dared to seek admission into the secondary school system overseen by the university.[2] Cambridge was designed to be an all-male institution, and Sedgwick, along with most other members of the faculty, viewed women's interest in higher education—particularly their interest in Cambridge—as no small threat to the natural order of things. It was not until 1871, more than six centuries after the university was founded, that women began to take classes. And even then, it was more of a reluctant nod than an embrace. Exactly five women were accepted that year; they lived in a little house on Regent Street.

From this foothold of change sprang Newnham College, founded in 1875 as the official school for women at Cambridge. As a 1919 fresher,

Cecilia—neither nasty nor a minx, but definitely forward—was one of six hundred Newnham undergraduate students. She quickly found out that although the winds at Cambridge might have changed, the atmosphere had not. The Victorian view of a woman—"that she was a creature born to please, whose personal individuality was strictly limited to, and by, that obligation"—still prevailed.[3]

Then there was the prevailing view of science: mathematics was for boys, botany was for girls. There were two "parts" to the study of the natural sciences at Cambridge. Three subjects had to be selected for Part One; and, if a Newnham student made it that far (Part One was rigorous and many students dropped out), she had to select one subject for Part Two. For Part One, a botany student would usually choose chemistry and zoology as the other two subjects.

Cecilia had been drawn to physics during her year at St. Paul's, and the field still enticed her. But choosing physics for Part One would have been both unconventional and risky—she could not change her mind or repeat it if she failed. So Cecilia hedged. She played by the rules and selected botany, but she insisted on taking the "unusual combination" of botany, physics, and chemistry. She was at Cambridge to become a scientist—maybe in botany, maybe not. She may have been socially shy, but she was forceful when it came to her education. Just as she had gotten her book on fungi, she got her way in course selection.[4]

From the beginning, however, Cecilia faced obstacles. As a scholarship student, she could not afford to buy all the required books. Her solution: before entering Cambridge, she went to the library, took Henry Dunkinfield Scott's *Studies in Fossil Botany* from the shelf, and transcribed it by hand into a notebook.[5]

Another problem was not as easily overcome. For someone who had already read a great deal, the introductory botany lectures were beyond familiar—they were boring. While peering at common algae under the microscope, she found some desmids, a type of microscopic green alga, that she could not readily identify. When she asked about desmids, the class demonstrator kept repeating that "they don't come into your course."

She concluded that desmids were "probably not his subject." She also felt cheated.[6]

Watching over Cecilia's growing disenchantment with flora and fauna was her great aunt Dorothea Pertz. Dora, as she was known, was a strong role model for Cecilia. She had been one of Newnham's earliest students and was a trained botanist. Aunt Dora thought some encouragement would be in order. She arranged to have Cecilia meet one of the great names in botany, William Bateson. Bateson did indeed provide encouragement, but not quite in the way Dora was expecting.

Bateson looked exactly like who he was: a cantankerous sixty-year-old man ready to do battle if you didn't agree with his thinking. Pictures of him capture his unruly head of hair, his only slightly more tamed moustache, and his permanent scowl. The scowl was there for a reason; he was taking on Charles Darwin himself.

Growing up in the 1870s, Bateson was routinely described as a mediocre student with little promise. He was labeled "a vague and aimless boy" by his headmaster at Rugby School. His classmates remembered that "he was never for half-measures or compromises" and "had little use for idle small talk." Later, moving from lab to lab, he was an irascible "dominating figure."[7]

Bateson enrolled at Cambridge in 1879. As Cecilia did forty years later, he concentrated on the natural sciences. He spent two summers studying *Balanoglossus*—more commonly known as the acorn worm—exploring its relationship to vertebrates, and gaining recognition in the established world of biology. Bateson soon turned on that world, however, and became an outspoken critic, claiming that traditional biologists were not rigorous enough in their analyses.

He did not believe that Darwin was correct in theorizing that evolution was caused by natural selection working on small variations. Instead, Bateson proposed that new species originated from large, discontinuous variation, and he championed the work of Gregor Mendel. In other words, evolutionary change was rapid and discontinuous, not slow and steady.

William Bateson

He devised a program of cross-breeding to demonstrate his theory and called the new field of study "genetics."[8]

Most biologists did not support Bateson's views. With research funds scarce, he turned to a vastly underutilized resource: the women of Newnham College. One of them was Dora. Bateson assigned her to study variation in *Veronica*, a plant with flowers of the same purplish hue as the bee orchis. He paid her poorly, but she was happy for the work.

Over time, Bateson's contributions were recognized. He left Cambridge in 1910 to become director of the John Innes Horticultural Institution in southeast London. It was in this plant breeding station that he gave Dora a gruff "yes" to her request that he meet her niece, Cecilia Payne.

Bateson showed Cecilia around the lab, explaining in great detail the various stages of his experiments. "It must be wonderful to do research," exclaimed nineteen-year-old Cecilia at the end of the tour.

Bateson stared at her, then barked, "It is *not* wonderful! It is exasperating!" He glanced with disdain at Wimbledon Church, which overlooked the institute's gardens. "We learn a lot of hymns here," he said acidly.

Cecilia was stunned, not so much by what he said as by how he said it. She told her mother so. "Poor man," responded Emma, "science is his religion."

"Well, it's mine, too," replied Cecilia. It was Emma's turn to be shocked.[9]

Bateson's caustic response may have brought Cecilia close to tears, but she claimed later that the man had done her a great service. This brief encounter with an already legendary scientist—his passion for rigor so obvious, his determination to understand so intense—was like a branding iron. Making discoveries did not result from just luck or dogged recording of observations; it required a painstaking, prolonged, systematic approach to problem-solving.

Cecilia came away emotionally shaken. She also came away with an appreciation for dedication, and with an even greater trial-by-fire determination. She and her fellow freshers were excited by the postwar possibilities for women. They had entered college in a much different era than the women of Newnham past. "The general feeling was one of enthusiasm and high spirits, and an impatience with old pre-war regulations," remembered a student who was one year ahead of Cecilia at Cambridge.[10]

This impatience for change made for tension. Off campus, women students were experimenting in unheard-of ways. Example: a good number of young officers, discharged at last from the British military, had returned to their studies at Cambridge. They brought not only "their strong arms" but also their motorcycles. "Every Sunday . . . the Newnham occupant of the sidecar would exchange places with the driver, and hoisting her skirts to mount astride the saddle, would learn to manipulate the beast, directed by the nervous owner."[11]

From sidecar to saddle was a small step; a bigger one was from saddle to smoking. Back in London, Cecilia's brother, Humfry, had introduced her to tobacco. He could not have known that it would become a habit that would shorten her life. She liked a cigarette, both for the taste as well as for what it represented. But Cambridge was not London, and increasingly "inside Newnham there developed a definite restlessness," recalled a student a few years older than Cecilia. When they were back home in London, students "were enjoying . . . a social freedom that made them resent the restrictions placed upon their freedom and their social intercourse at college." It was particularly annoying "to be still solemnly adjured that a cigarette was an offense against decorum, and that to be noticeable was a crime." Thus, when Newnhamites had nicotine cravings, reminisced one student, "we committed the unheard-of crime of going out to the field just beyond Newnham grounds, and, seated with our backs to the hedge, indulging in a cigarette!"[12]

Nor apparently was a surreptitious smoke limited to undergraduates chafing at custom. Blanche Athena (everyone called her "B. A.") Clough—niece of Newnham's founder Anne Jemima Clough—was now the college principal. "It was rumoured, to everybody's delight, that B. A. herself used to take a cab out to a certain spot on the Grantchester Road, and enjoy a quiet cigarette in the fields."[13]

Cambridge was in every way a severe surrogate parent. There were rules, and decorum had to be maintained. Students under the majority age of twenty-one were deemed to be children, and after sunset, Cecilia and all undergraduate students, both male and female, were required to wear gowns. Because she was under twenty-one, Cecilia was required to be back inside Newnham's walls by 10 P.M. Once the doors were locked, there was no sneaking in, or out. To enforce the policy, the walls around the various colleges had shards of broken glass embedded along the top.

Patrolling at night, ever on the lookout for excessive adolescent enthusiasm, were the "bulldogs." Their official title was proctor—burly
men dressed in buckled shoes, spats, and bowler hats. If a bulldog col-
lared a student after ten o'clock, his or her tutor would receive a disci-
plinary note the next morning.

It was a closed system of enforced primness in which permission was
the order of the day. Or night. Cecilia could beg for a rare night out; if
she received it, she still had to sign in at the Porters' Lodge just inside
the gate. "One wasn't allowed out much in the evenings," recalled one
student a year ahead of Cecilia, "and if you were going to a dance club
once a week you were lucky if you got permission to go out again the
same week, and only for something like a concert or a really good play—
something of that sort."[14]

Apparently, the event that took place on Tuesday December 2, 1919,
qualified as "something of that sort." It was not a concert or a really good
play—it was a lecture, and it was all anyone at Cambridge could talk
about. On this chilly night in the winter of Cecilia's first year, the astronomer Arthur Eddington was going to report on how his recent expedition to Africa to observe a solar eclipse had provided experimental
confirmation of Einstein's general theory of relativity.

Eddington was only thirty-seven years old, but he had already succeeded Sir George Darwin (son of Charles Darwin) as Plumian professor
of astronomy and experimental philosophy at Cambridge, the most prestigious astronomy chair in Britain. He was now director of the university's
observatory. With his beautifully tailored three-piece suit, white starched
high-collar shirt, perfectly knotted silk tie, and a thin gold pince-nez
perched lightly on his nose, he was a very handsome man. Just the hint
of a smile was his way of signaling that he knew more about whatever a
student might say than he or she did.

The venue this evening was Trinity College's dining hall, where "rats
came out and gnawed remains left on the floor!"[15] The crowd outside, five
hundred strong, was pushing and shoving to get in. Unmindful of the
cold and damp, they were mostly men. Only four of the hottest tickets in

Sir Arthur Eddington

town had been allotted to Newnham College. When one of the four lucky students fell ill, she gave her ticket to Cecilia. Ticket in hand, Cecilia was not about to let rats or regulations deter her. She and the three other women pushed their way through the crowd and found four seats together.

The student who gave Cecilia her ticket surely knew that her class-mate had an interest in the topic. What she could not possibly have known was that her simple act of thoughtfulness would lead to a turning point in Cecilia's life. After the lecture, Cecilia later recalled, "for three nights, I think, I did not sleep. My world had been so shaken that I experienced something like a nervous breakdown."[16] She had had an instant recognition of a life's calling that years later would lead to unlocking one of the great mysteries of the universe.

7

One of the quirkier Cambridge traditions was the Great Court Run, immortalized in the film *Chariots of Fire*. Every year, Trinity College students attempted to run around the Great Court, approximately 320 meters, within the time it took for the college church clock to strike the hour of twelve noon. And for the beat-the-clock race in 1910, the student runners were furious with Edwin Turner Cottingham.

Cecilia never met Cottingham, and she probably did not even know who he was. But in his low-profile, humble way, Edwin was related to the group of scientists who would inspire Cecilia to reach her own level of greatness—for Edwin Cottingham would make Arthur Eddington's discoveries possible, and Arthur Eddington in turn would make possible those of Cecilia Payne.

Cottingham was born in 1869 in the village of Ringstead in northeast England. As a boy, he was fascinated by all forms of mechanical instruments. He was very good at taking things apart, and he was very good at putting them back together—they always seemed to run better afterwards. He had no manual or instructions, just a natural intuition.[1]

Clocks gradually captured his complete attention, and at a young age he was a successful self-employed watchmaker. Once he had mastered the mechanics, precision became his passion. He paid the local post office £5 a year to transmit by wire the precise 10 A.M. signal to his workshop from the Greenwich Royal Observatory. He was granted a wireless license so he could receive a time signal from the Eiffel Tower.

And so it was to forty-year-old Edwin Cottingham that Cambridge University turned when the Trinity College church clock needed cleaning in 1910. And did he clean it!—drawing not only the ire of runners but the eye of the press. "Mr. Edwin Cottingham played at least a small part in deranging records," wrote *The Observer*. "In tending the clock of Trinity Church, Cambridge, he speeded it up slightly, spoiling the sport of the undergraduates."[2] It would be seventeen years before a runner finally succeeded in reaching the finish line before the chimes struck twelve.

Ten years later, it was to Edwin Cottingham that the Royal Astronomical Society (RAS) turned for ensuring precision in what would be one of the most famous experiments in scientific history. Edwin's obsession with clocks had led him into astronomy, and he had been elected a fellow of the RAS in 1905. In the spring of 1919, the year Cecilia went to Cambridge, he was asked to meet with the physicist and astronomer Arthur Eddington to map out plans for an expedition to view a solar eclipse. Edwin was thrilled. He knew his clockwork precision was prized. But he would probably never have guessed that a historian would someday observe that "with Edwin Cottingham, we see the move from Victorian science into the world of Einstein."[3]

Of all the teachers and lecturers in Cecilia's life, none would affect her as much as Arthur Eddington. From an early age, Eddington showed himself to be a brilliant scientist. He was the first student in Cambridge history to achieve the ultimate accolade, senior wrangler in mathematics, after only two years of study. But with his impeccable dress, his audience-pleasing lecture style, his penchant for spirited physical exploits such as glissading—hurtling down steep grassy slopes at the expense of the seat of his pants—there was a showman side to Eddington as well.

Eddington was supremely self-assured. This is how he opens chapter 11 of his book *Philosophy of Physical Science*: "I believe there are 15,747,724, 136,275,002,577,605,653,961,181,555,468,044,717,914,527,116,709,366, 231,425,076,185,631,031,296 protons in the universe, and the same number of electrons." Although he remarked that he did not rank this estimate as one of his strongest convictions, he added, "I am, however, strongly convinced that, if I have got the number wrong, it is just a silly mistake."[4]

In 1915 Albert Einstein had published a paper entitled "The General Theory of Relativity." It was controversial both scientifically and politically. England and Germany were at war, and what this German scientist was claiming was that the British scientist Sir Isaac Newton had been wrong about gravity. Gravity, claimed Einstein, was not a force that acted instantaneously at a distance, but rather was a warping of space-time that could affect even light itself. Einstein hypothesized that light would bend as it passed by a large mass such as the sun. Eddington was one of the few people who could understand the math behind this theory. If he could find a way to test the theory, he would be forever linked with a historical breakthrough.

Frank Dyson, Britain's astronomer royal, agreed with Eddington that a test would be critical, and he had proposed a clever way to do it: shield the sun so that it would be possible to see the light from stars that appeared to be close to the sun (that is, distant stars whose light just skimmed by the edge of the sun on the way toward the earth). If light from those stars did indeed bend as it passed close by the sun, then those same stars would appear to an observer to shift position ever so slightly. But how to shield the sun? By taking advantage of a solar eclipse. And there was an eclipse fast approaching, on May 29, 1919.

There had been several unsuccessful attempts to use a solar eclipse to verify the bending of starlight. This time, however, the timing was fortuitous, and so were the conditions: the sun would be at the center of Hyades, a very bright field of stars, perfect for measuring light deflection. But to observe it, the astronomers would have to travel to the remote island of Príncipe, off the coast of West Africa.

Eddington knew that the entire mission depended on accuracy. Newton had believed that light had a small amount of mass, so gravity would bend it slightly. According to Einstein's calculations, based on his new theory of gravity, the warping of space would cause light to bend *twice* as much as Newton had predicted. Nonetheless, it would still be difficult to measure: "A ray of light nicking the edge of the sun, for example, would bend a minuscule 1.75 arcseconds—the angle made by a right triangle 1 inch high and 1.9 *miles* long."[5]

Work immediately began on dismantling telescopes and other astronomical devices from observatories in Cambridge and Oxford and preparing them for shipment. The task to come was daunting; the coelostats—mirrors used in solar observations—would have to be perfectly driven by small clockwork devices in order to counteract the earth's rotation and keep the starlight focused on telescopic camera lenses for long intervals of time. And the observations would be made out in the field, not in a comfortable, controlled observatory dome. Even Arthur Eddington would need help. And who better than the clockmaker who had so agitated the runners at Trinity College with his penchant for pinpoint precision?

When Cottingham finally met with Dyson and Eddington, the observatory equipment was already packed and ready to go. Dyson took great pains to impress on Cottingham how vital his responsibility was for calibrating the telescopes. He "told the clockmaker that there were three theoretically plausible results: no deflection; half deflection, which would show that light had mass, and vindicate Newton; and full deflection, which would vindicate Einstein. Gathering that the greater the deflection the more theoretically exciting and novel the result, Cottingham asked what would happen if they obtained twice the Einstein deflection."[6]

"'Then,' said Dyson, 'Eddington will go mad, and you will have to come home alone.'"[7]

Eddington and Cottingham sailed from Liverpool on March 8, 1919, bound for Madeira on the southwest coast of Portugal, loaded down with "telescopes, crates, canvas, mirrors, cigarettes, two metronomes, no doubt plenty of tea, and other essential items."[8] They arrived on Príncipe on April 29, 1919, a month before the eclipse was due. It was a densely forested island with surf smashing against the base of five-hundred-foot-high cliffs and clouds ringing inland mountains a half mile high. They had to rely on local porters to carry the equipment to a clearing. "Rain, mosquitoes, and quinine became the daily regimen. Eddington and Cottingham built waterproof huts for the equipment, with the help of laborers from a local plantation. They were forced to work under mosquito netting and at least once helped hunt monkeys that had been interfering with their equipment."[9]

The dawn of the eclipse: no monkeys, but rain. As Eddington recorded in his eclipse journal, "[in] the morning there was a very heavy thunderstorm . . . a remarkable occurrence at that time of year."[10] The eclipse would begin at 5 seconds after 2:13 P.M.; Cottingham had the metronome ready to tick off the seconds during the eclipse. When the crescent of the sun disappeared, plunging the forest into near total darkness, he yelled, "Go!" They had traveled thousands of miles, hauled delicate equipment overland, and braved monkeys and mosquitoes in order to record a moment in time that would last just five minutes.

They worked furiously, exposing sixteen photographic plates. The clouds had parted enough for the team to capture fuzzy but useable starlight just above the sun's corona. Eddington cabled Dyson: THROUGH CLOUDS STOP HOPEFUL STOP EDDINGTON. As he worked on developing the plates, Eddington's confidence grew. "Three days after the eclipse, as the last lines of the calculation were reached, I knew that Einstein's theory had stood the test and the new outlook of scientific thought must prevail. Cottingham did not have to go home alone."[11]

When Eddington reported his findings in November of that year, the rest of the world took note. The *London Times* was more or less restrained in its coverage the next morning; not so the *New York Times*. On short notice, the New York desk had assigned the story to its golf correspondent, Henry Crouch. "He would have been the first to admit that he was very much not an authority on the mathematics of four-dimensional space-time. Crouch did, however, work out that something extraordinary had occurred, and his enthusiasm was transmitted to the *New York Times* headline writers."[12]

On November 10, 1919, the paper proclaimed:

LIGHTS ALL ASKEW IN THE HEAVENS
Men of Science More or Less Agog Over Results
of Eclipse Observations

EINSTEIN THEORY TRIUMPHS
Stars Not Where They Seemed or Were Calculated
to be, but Nobody Need Worry

Eddington had made Albert Einstein a global celebrity, and he was quick to link Einstein's cerebral thought experiments with his own physical exploits. He wrote to Einstein, "I have been kept very busy lecturing and writing on your theory" and noted that at a recent Cambridge Philosophical Society lecture, "hundreds were turned away unable to get near the room."[13]

As Cecilia listened to Eddington recount his scientific exploits, there was a milling mob of students and teachers outside the dining hall, hoping to get a glimpse of the man who made the headlines. To Eddington, Cecilia was nothing more than an anonymous face in a Cambridge audience; that would soon change. To Cecilia, Eddington was like a prophet, reaching out to her, encouraging her to acknowledge her calling. "The result was a complete transformation of my world picture," she recalled later. "The experience was so acute, so personal."[14]

Cecilia did not linger. She described the effect of Eddington's speech on her as a "thunderclap." She raced back to her dorm room and transcribed Eddington's entire lecture—word for word—into a notebook.

Despite Eddington's siren call, she kept at her botany studies—sitting through lectures, peering through the microscope, using the sketching skills she had learned from her mother, Emma, to draw plant cells. The force keeping her, for the moment, focused on botany was the result of yet another encounter, with Agnes Arber.

Agnes Robertson Arber, born in 1879, was like a hybrid between Cecilia and Emma. Like Cecilia, she had attended a science-oriented high school, North London Collegiate School for Girls, where the emphasis was on academics, not traditional homemaking, and where she won numerous awards and a scholarship to study botany. Like Cecilia, she could not have cared less about social niceties, and summer vacations were for working in the lab. She, too, was drawn to Cambridge. She completed the work for a natural sciences degree from Newnham in 1902 and relocated there when she married the paleobotanist Edward Alexander Newell Arber in 1909.[15]

Like Emma, though, she was a single mother. Agnes had a daughter, Muriel Agnes Arber, born in 1913. Her husband died in 1918, the year before Cecilia entered Newnham, and she never remarried.

Agnes channeled her grief into her work—ferocious, relentless, nonstop research. At the time, no woman at Cambridge University, whether student or faculty, could take out books from the university library unless a man signed for them. But Agnes was not deterred. She produced more than fifty papers and wrote three books on plant morphology that ended up in that very library. When her research lab closed in 1927, she appealed to A. C. Seward, the head of Cambridge's botany department, for use of the school's facilities. Seward turned her down. So she built a laboratory in her home and continued her research alone, emerging in 1946 to become the first woman botanist member of the Royal Society. She lived on such a meager income that when she died in 1960, there was still no electric service in her home.

Agnes was assigned to be Cecilia's botany tutor, and she was far more sympathetic than the men tutors. Nonetheless, she was not one to offer

Agnes Arber

false encouragement. When Cecilia eagerly submitted a carefully re-
searched essay on the evolution of root structures, Agnes deemed it neither original nor significant. To an exuberant first-year student, it was
devastating criticism, and it further fueled Cecilia's questioning of
whether she was made to be a botanist.

Agnes's critical eye may have been turning Cecilia away from plants,
but Cecilia was still learning from her. In her leisure time, Agnes wrote
poetry, and her morphology papers were as lyrical and literary as they
were technical. If you were an astute student, you perceived that scientific
precision could be artistic as well. And Cecilia was a very astute student.

But Cecilia, a believer in Darwinism, was impatient with Arber's traditional approach to botany; she was excited by natural selection and bio-
chemical discoveries. "The spirit of a new, rational biology was in the

air," she wrote later, "and the old picture seemed as démodé as the Ptolemaic system."[16]

Cecilia was ambitious. She had long wanted to come up with big new ideas. She once had told her teacher Ivy Pendlebury of her aspiration to do research. "Why do you want so much to do research?" Ivy asked.

"It will be so wonderful to make new theories," was Cecilia's instant answer.

Ivy responded by pointing out that very few people make new theories. Most scientific work, she explained consisted of making accurate observations.[17]

Ivy was not only an intelligent science teacher; she was also wise. But in the same way that Cecilia was not "just like other girls," she would not be like other scientists. For sure, Cecilia would observe and measure— she would do all the routine daily activities of a traditional scientist. But she was also quietly competitive. Soon she would fulfill her own expectations and surprise even Ivy.

Before she had a chance to come up with new theories, though, she faced some setbacks. Ever an avid collector, and eager to question existing classifications, Cecilia came upon a "remarkable rose" while hiking along the cliffs of Cornwall. She brought her painstakingly drawn rendering to the herbarium of London's Natural History Museum, completely convinced that she had discovered an entirely new species. The curator was an older man who had seen his fair share of student breakthroughs. "I suppose you tagged the bush?" he asked wearily. Cecilia was "crushed" at the realization of how inexperienced she was.[18]

More self-questioning. Cecilia viewed her mistakes as indications that perhaps she was not meant to be a scientist. The truth is, she was still reluctant to acknowledge that botany was not for her. She claimed that she wanted "to turn from an empirical science to one in which one knew what one was talking about."[19] She would come to realize, however, that she was rationalizing. Years later would she look back and acknowledge that probing the structure and makeup of the atomic nucleus is just as challenging as trying to understand the structure of organic molecules.

She chafed. She knew there was a place just across the river—quiet, nondescript, tucked in among identical-looking Gothic buildings— where secrets of the atom, sought by scientists for centuries, were suddenly being revealed. "The lure of the Cavendish Laboratory was irresistible," she recalled later, and she longed to be a part of it with nothing short of passion. "I was eager to be off with the old love and to embrace the new, my beloved physics."[20]

8

Eddington's mesmerizing performance and the lure of the Cavendish Laboratory proved to be too much. After her first year at Cambridge, Cecilia was done with botany and "dedicated to physical science, forever." (Halfway into that year she had alerted the college administrators that, in the parlance of Cambridge, she was going to "change her shop" and read physics.)[1]

But reading physics meant going to the Cavendish Lab for class, and the building was across the river in the middle of town. The most efficient way to get there was by bicycle, which involved no small amount of risk. "Biking [is] a fearful job there," recalled one of Cecilia's contemporaries, "especially in Silver Street and Petty Cury."[2]

Petty Cury was an old winding street, packed with shops and inns. Betty Leaf, Cecilia's close friend, would confirm just how "fearful" this slice of Cambridge could be. It was a wonder they didn't get run over. Cecilia's "absorption in abstract questions is sometimes fraught with danger," Betty wrote. "I have known her [to] suddenly descend from her

bicycle amid a Saturday crowd in Petty Cury, and unaware of the whirl-
pool of vehicles, abruptly enquire 'What's the good of thinking?'"[3]

Ever since Newnham's earliest days, pedaling hard on two wheels has
been the best way to get around. "I well remember the cumbrous skirts
covering trouserlegs firmly secured to ankles with broad black elastic,"
recalled a young Newnhamite about her college days in the 1890s.
Walking between classes would have taken too long. "Most lectures took
place in the morning," recalled Enid Mary Russell, who graduated a year
before Cecilia, "and we hastened from one to another in the traditional
fashion, on bicycles, our notebooks in the basket on the handlebars, and
in winter a muffler round our necks."[4]

It was a familiar sight each morning: a parade of students on bicycles
threading their way over bridges and along narrow Cambridge streets.
"You haven't lived until you have done your weekly grocery shopping
on a bicycle in the pouring rain," said Virginia Trimble, an astronomer
who went to Cambridge in the 1960s.[5] Most of the women played by the
dress code rules—blend in, be "mouselike"—although there was the occasional outlier. Enid Russell, for example, managed to "set off my
khaki flannel suit with a muffler in pegeon's-neck [sic] colours knitted by
Mother. This would fly out bravely behind, until one day the end caught
on a bicycle going in the opposite direction with the most disastrously
undignified results."[6]

As in other arenas, women had to abide by different rules than men.
There was a toll, so to speak, in order to gain access to town: a hat. Size,
color, style, age—it didn't really matter as long as the head was covered,
for in early twentieth-century Cambridge, women "were forbidden to go
to lectures, or into town at all, without hats." They ranged from old-
fashioned straw boaters, to "small ones tipped up at the back," to "large
cart-wheels." Or a woman might don a "wide flowery straw hat," perhaps with a "chocolate-box ribbon around it."[7]

Given Betty Leaf's family background, acquiring a selection of costly,
stylish hats would not have been a problem for her. But for Cecilia, a hat
was not a minor expense. Fortunately, there were stores like the one at

the corner of Market Passage and Sidney Street, described by a Newnham student as "a cheap draper's shop that sold all sorts of things. One day it had in the window a lot of hats, old-fashioned hats even in those days, and they were labelled 'Absolutely the limit' . . . pepper-and-salt straws— the sailor type of things."[8]

Cecilia could be hatless at first as she wheeled her bike out the Newnham gate. But within a few blocks, her head needed to be covered. "There was a definite demarcation point, somewhere near the Newnham end of the Silver Street bridge, at which the town was deemed to have begun, and there we must stop and cover ourselves decently before proceeding further."[9]

Decently covered, she would pedal quickly down Sidgwick Avenue toward the river, making good time for the several blocks to Queens Road because there were no traffic lights at the time. Not to say she could be careless—horse droppings littered the streets, "and the street sweepers were constantly passing up and down with their brush and shovel, pushing a small hand-cart."[10]

A minute later she would cross over the punts on the River Cam, no doubt glancing to her left at the Mathematical Bridge, a footbridge famous even then. Constructed entirely of wood, it was (and still is) a self-supporting arc using only straight pieces of timber. The tangent timbers were compressed against each other so that the radial timbers fitted together with little bending stress, all simply and elegantly illustrating, in true Cambridge fashion, the forces of arched construction.

She would continue down Silver Street to the T intersection at Trumpington Street. A right and then a quick left would put her on a biker's shortcut, Botolph Lane. Too narrow for anything but bicycles, Botolph (named for St. Botolph's Parish Church on the corner) was a block-long collection of two-story houses and shops. It ended at Free School Lane. All in all, a ten-minute ride, though in the middle of a damp English winter, it could seem an eternity.

Just up the block on Free School Lane, there it was—the gothic imposing entrance to the famed Cavendish Lab. For a twenty-year-old woman bicyclist, it must have been intimidating. The walls, composed of massive interlocking stone blocks, were eighteen inches thick. The windows were framed with specially constructed wooden shutters to provide the darkness required for electromagnetic experiments that produced only dim flashes of light.

It's hard to believe, but before the Cavendish Lab was built, a specially designed experimental physics laboratory—a gathering of working physicists and a collection of instruments all under one roof—had never existed. Up to then, experiments in physics were conducted either by wealthy amateurs in their home laboratories or by university students in their dorm rooms. The idea for the lab emerged in 1869 when the Cambridge Senate decided to take up the question of whether the university should expand beyond the teaching of theory and pursue practical training of scientists and engineers.

There were obstacles. One was heresy. Experimentation was not what Cambridge intellectuals did. Mathematics ruled—it was commonly referred to as "the vampire of the Cambridge schools," sucking up resources and leaving little for the arts and the natural sciences. Students in the Mathematical Tripos used their minds; they decidedly did not work with their hands. As the Cambridge mathematician Isaac Todhunter put it, "Experimentation is unnecessary for the student. The student should be prepared to accept whatever the master told him."[11]

There was also the small matter of funding a radical new institution. Although the Senate voted to proceed, the university's finances were not robust. The chancellor of the university at the time was William Cavendish. He was also the Seventh Duke of Devonshire—which meant he was rich. After eighteen months of no activity, William wrote to the university and offered to pay the full cost of construction—£6,300—if the university would fund the hiring of the first director. The university promptly agreed to name the new institution the Cavendish Lab and hired James Clerk Maxwell to oversee the design.

Maxwell, like Cecilia, was a born scientist. In 1841, at age ten, he was sent to Scotland's prestigious Edinburgh Academy, where he won awards in biology, math, even poetry. His first scientific paper, *Oval Curves*, was published by the Royal Society—when he was fourteen.

With bushy sideburns sprouting left and right, and with an equally bushy beard, his gentle expressive face was at the center of a halo of hair. But the quiet demeanor and playful sense of humor belied a personal drive. He believed that progress in physics, and thus the opportunity to make discoveries, required one thing above all: measurement. Theories were useless if they could not be tested. And in order to create a center of measurement he knew he would need stuff—gauges, lenses, prisms, meters—and a place to house it all.[12]

Maxwell also knew that a hands-on laboratory within a singularly focused mathematics school would have to maintain a low profile. In fact, it would have to be hidden. And often the best place to hide something is in plain sight. Thus the Cavendish Lab—one of the birthplaces of modern physics, where the secrets of the physical world would be unlocked, where young scientists eventually would conduct experiments in atomic fission—was not located at a safe distance somewhere out in the countryside, but right in the middle of town.

A cobblestone entrance leading to a thick wooden gate, a pointed stone archway inscribed with Psalm 111, verse 2 (*Magna opera Domini exquisita in omnes voluntates ejus*, "The works of the Lord are great; sought out of all them that have the pleasure therein")—it deliberately did not look like what today would be a sophisticated center of high technology; it looked like a medieval fortress.

Unless one looked closely. The platforms beneath the windows, for example, did not hold flower boxes. Maxwell figured that if the lab's scientists were to be able to measure electricity and magnetism and heat with precision, they would need steady light for prisms and lenses and cameras. He replaced flower boxes with helioscopes, clock-driven mirrors that followed the motion of the sun and drove bright light into the windows.[13]

It was subterfuge at its finest. One can imagine Maxwell's delight as he wrote to his colleague, and future successor, Lord Rayleigh: "If we succeed too well, and corrupt the minds of youth till they observe vibrations and deflections . . . we may bring the whole University and all the parents about our ears."[14]

What Maxwell succeeded in building was in actuality a high-tech startup. It had all the characteristics of entrepreneurialism: venture capital, cost overruns, egos, breakthroughs. It was very much the Silicon Valley of its time—it featured open spaces with intensely focused engineers bent over worktables cluttered with gadgets. But unlike many of today's ventures, it was not lavish. This was a lab devoted to experiments—the discipline after all was called "experimental physics"—and funds were to be used for equipment, not lifestyle. There wasn't a dress code per se, but there was an atmosphere, a belief that disciplined work required disciplined attire. Everyone wore a suit, a starched high-collared shirt, and a necktie.

There was a distinct Dickensian air to the place, "with its cobbled courtyard and its archways and massive oak gates, locked and unlocked religiously twice a day with much clanking of iron keys in locks."[15] And as the lab matured, there were whisperings and rumors as to what was going on inside those thick stone walls. It certainly wasn't theorizing, so what was it? Myths took hold. One was that the mathematicians turned physicists, who at first didn't know a thing about measurement, were amateur scientists conducting experiments with "sealing wax and string." They were not. Another was that the equipment was crude, homemade. It is true that the vacuum tubes preserved today in the Cavendish Museum, with their small glass nipples holding delicate wires, look like chemistry-set toys. They are not. The glass-blowing alone was extraordinarily sophisticated for its time.

For intellectually gifted mathematics students, this was a very different kind of instruction than what they were used to. The Cavendish was an immediate hit, quickly attracting graduates from other universities. The first arrivals were not readily accepted, however, because they had no formal student status—officially, Cambridge did not grant research degrees in this new field of experimental physics; that would not come for another twenty years.

Nonetheless, word spread, and a growing group of future Nobel laureates flocked to Maxwell's Victorian vision, drawn by its cultivated nose-thumbing at the traditional intellectual class. Bent over their research tables, in their tweed suits, they took the first steps toward detecting isotopes, developing x-ray crystallography, and perfecting the splitting of the atom. It is a miracle they did not blow themselves up. Even today there is an inspector who has spent an entire career checking for toxicity at the site of the original lab.[16]

Almost everything in Cambridge was narrow—especially the stairwells inside the lab. As Cecilia climbed and descended the wooden stairs, like everyone else, she would have to turn slightly sideways to let others pass. The result was that, both up and down, Cecilia would have rubbed shoulders with many current and future Nobel laureates.

There was Alexander Wood, a Scottish scientist who wove acoustics into his lectures and experiments. Wood studied under Lord Kelvin at the University of Glasgow, where he acquired two pronounced characteristics: his habit of praying before each lecture, and his thick brogue. Cecilia considered his voice to be golden, portraying his distinctive burr in her memoir: "'I canna believe,' he declaimed with his Paisley accent, 'that the Univairse is a collosal [sic] prractical jooke on the parrt of the Creatorr.'"[17]

There was Francis William Aston, an Englishman who was recruited to the lab for his ability to fuse chemistry with physics. Aston worked alone, hunched over his worktable, examining his blown-glass tubes, brass fittings, coiled wires. He believed wholeheartedly in the Cavendish credo: "Now, what would happen if we tried . . . this."

Aston personally built the batteries for huge electromagnetic plates so that the experimental voltage would be steady to one part in a hundred thousand. He designed and constructed an enormous magnet that produced a force field thirty thousand times stronger than that of the earth.

In 1919, the year Cecilia arrived at Cambridge, Aston produced one of the world's best mass spectrometers, an instrument that determines the mass of a single particle or stream of particles (such as ionized atoms or molecules) by bending their paths with electric and magnetic fields. (Today, the power of mass spectrometry is used in applications ranging from analyzing the breath of anesthetized patients to uncovering drug use in athletes.) For his work, Aston won the Nobel Prize in Chemistry in 1922. He often mused to his colleagues that perhaps some day it would be possible to generate nuclear power.

And then there was the great Danish physicist Niels Bohr, a soft-spoken man Cecilia strained to understand, whose "discourse was rendered almost incomprehensible by his accent. There were endless references to what I recorded as 'soup groups,' only later emended to 'sub-groups.'"[18]

The cutting edge of physics at the time was the "Bohr atom," and it was Bohr himself who visited Cambridge to describe his theory of atomic structure in a series of guest lectures at the Cavendish Lab. He was only in his thirties when he strode into the lecture hall and, as Cecilia listened, "cut the ground from under the majority of physicists."[19]

About 450 BCE, the Greek philosopher Democritus proposed that all matter was composed of particles called atoms. Democritus theorized that atoms were distinguished from each other by their sizes and shapes; some came with hooks and eyes, others with balls and sockets. That's about as far as anyone got until the early part of the nineteenth century, when the English physicist William Prout hypothesized that the atomic weight of every element in nature was a multiple of the weight of the same fundamental particle. That particle seemed to have the same mass as a hydrogen atom; Prout named it the "protyle." He wasn't quite correct. It was the physicist Ernest Rutherford who discovered in 1911 that Prout's particle was in fact the nucleus of the hydrogen atom and had a positive electric charge, while a negatively charged electron orbited around it.

Rutherford clearly had Prout in mind though when, during a meeting of the British Association for the Advancement of Science in 1920, he was

Niels Bohr

asked to name the particle he had discovered. "Rutherford promptly offered two, *proton* and *prouton*. Both names were apt. Proton carried the Greek root for *first*, suggesting a primary form of matter. *Prouton* was even more explicit. It conjured up Prout's Hypothesis."[20] *Proton* eventually won out.

It had been the Cavendish Lab that had jump-started progress in understanding the atom. Decades earlier, under the leadership of J. J. Thomson, the lab had developed the "plum pudding" model: atoms were described as being "blobs of positively charged pudding studded with negatively charged plums." The problem with the pudding was that it was half-baked. Rutherford maintained that atoms actually consisted of a dense positively charged core with electron plums whizzing around at a distance. But Rutherford's atomic model also posed problems. Per classical physics, opposites attract—so why didn't negative electrons just spiral down into the positively charged nucleus?[21]

Then along came Bohr with an answer as mysterious as the mystery. He told Cecilia and the other physics students in the Cavendish audience to brace themselves. Electrons did indeed orbit a nucleus; but an electrostatic force was what held it all in place, not gravity. And the position of the electrons was not random. From work he had completed in 1913, Bohr described his theory that electrons could occupy only certain specific paths, orbiting at precise fixed distances from the nucleus. They could move from one fixed orbit to another, but they couldn't go halfway—they had to make a *quantum leap*. And when they dropped down from a higher energy orbit to a lower one, they would radiate energy.

All Bohr was asking was that Cavendish Lab physicists abandon decades of classical physics and convert to his world of quantum mechanics. It was brilliant; he was fusing physics with the quantum concept and then applying the result to the problem of understanding atomic and molecular structures. The listeners reacted along the lines of the wider world: some bought in wholeheartedly; some were baffled beyond words. Many agreed with a jest attributed to William Lawrence Bragg, a physicist and future director of the Cavendish Lab: "God ran electromagnetics on Monday, Wednesday and Friday by wave theory, and the devil ran it on Tuesday, Thursday and Saturday by quantum theory."[22]

Like all good speakers, Bohr knew his audience. The Cavendish crowd was made up of experimenters, and he told them the one thing he knew would hook them: his model was measurable. When an element is heated or subjected to an electric current, its electrons become "excited" and emit light in a distinctive color. So, Bohr illustrated his lecture with spectra, slides displaying an excited element's particular pattern of wavelengths. He showed how the spectrum for hydrogen, for example, could be linked to the "quantized" behavior of its electrons. The demonstration was simple, straightforward, elegant. Bohr received the 1922 Nobel Prize in Physics.

George Frederick Charles Searle

Not surprisingly, Cecilia adored his lectures. Like Cecilia, Bohr was well read, and he used literary references to help explain scientific truths. Like Cecilia, he had a relentless penchant for understanding—as a child he would fix instruments, particularly clocks, that needed repair. Like Cecilia, he "concluded that religion as taught could not withstand scrutiny in the context of logic and science."[23] In fact, Bohr looked upon classical physics the same way he saw religion, believing that "ordinary mechanics represented the truths of the microworld no better than conventional religious beliefs accorded with the meaning of life."[24]

Cecilia no doubt gazed at Bohr's slides and used her childhood lessons in illustrating to record his spectrograms. To Cecilia, Bohr's explanation for how a hydrogen atom could produce a recognizable pattern of lines was so direct, so simple, so understandable. She filed away Bohr's hydrogen spectral lines in a mental folder—recognizable and readily recallable when the need would arise.

And finally there was George Frederick Charles Searle. Searle was a researcher in electromagnetism, but he was best known as the head of the undergraduate labs. With a thick beard and explosive temper, for fifty-five years he supervised wave after wave of budding university physicists. He had his own parental style. When a student made any kind of mistake, he or she was told to go stand in the corner.

He too prized measurement above all else, and he was a stickler about it. "He had no patience with the women students," Cecilia recalled later. "He said they disturbed the magnetic equipment, and more than once I heard him shout, 'Go take off your corsets!' for most girls wore these garments then, and steel was beginning to replace whalebone as a stiffening agent."[25]

Searle may have been eccentric, but beneath the short temper and the brusque manner he was a dedicated scientist. He spent hours training Cecilia and her classmates to be skillful in measuring, precisely, all forms of scientific data.

What she did not know, however, was that he was intensely involved in another form of "science." He had contracted some form of disease during World War I, was unexpectedly cured, and promptly became a Christian Scientist. When Cecilia approached him for help in keeping up with the heavy work schedule, Searle assured her, "There's nothing wrong with your mind. It's your *soul* that needs attention." He then carried her off on the spot to a Christian Science healer. Cecilia endured the session in silence, thanked the woman, and never sought Searle's help again.[26]

9

At first, Cecilia's lectures and laboratory work at the Cavendish Laboratory were steeped in classical physics: mechanics, electromagnetic theory, thermodynamics. But the lab was packed with future Nobel laureates—a critical mass of scientific imagination and ideas. That many minds working in a contained, elbow-to-elbow environment not only generated new theories but also provided the means to verify them. The subject of radioactivity dominated the lab, and Bohr's radically new quantum theory injected excitement into physics classrooms.

Eddington's words, though, were still fresh in Cecilia's memory. The stars called, but to get to them, she faced a hurdle: she could not transfer to astronomy because technically it was a branch of the mathematics department, not physics. She did not know it at the time, but the inability to transfer would prove to be extremely fortunate, both for her and for astronomy. Astrophysics—the use of physics to understand the composition of stars and other celestial bodies—was emerging as a discipline, and by staying in the physics department, she would be perfectly positioned to enter the new field.

Though she couldn't join the astronomy department, she could drop in on lectures, and she "fell on them with avidity. . . . The rest of my time at Cambridge was to be devoted to physics, with all the astronomy I could pick up on the side."[1]

For someone as driven to understand as Cecilia, one can only imagine the euphoria of finding a life's calling. She would no doubt have laughed if told that her excitement in learning about the heavens put her in the company of Copernicus, Tycho Brahe, Galileo, and Kepler. With all the resources available to her at Cambridge, it did not take long for her to begin reading all the astronomical books she could get hold of.

She quickly discovered Henri Poincaré. Described as a "monster of mathematics," Poincaré was one of the greatest of European mathematicians. Working in both pure and applied math, he would lay the foundation for modern chaos theory. Cecilia didn't just browse his books; she attacked them. "I remember finding [Poincaré's] *Science and hypothesis* in the library at Newnham, and sitting on the floor and reading it from cover to cover on the spot."[2]

She also discovered that Newnham College had an observatory. Three years earlier, another Newnham fresher named Dora Clarkson Lawe learned that there was an Astronomical Society, became a member, and was able to secure the key to the observatory. She waited for nightfall: "Dropping out of a ground-floor window at dead of night (after suitable conspiratorial arrangements with the occupant of the room to let me in again in the small hours) I taught myself in an amateurish way to open up the shutter of the observatory roof, work the ratchet which swung it around and focus the telescope lens so as to observe the moon."[3]

The observatory was located at the far end of the expansive Newnham hockey field. The great lawn was treeless; there was nothing to block the wind. For would-be astronomer Dora Lawe, it proved too much. As she snuck out, alone, her footsteps in the grass the only sound, "enthusiasm for these unlawful nocturnal expeditions waxed and waned, like the object of my observations, and was finally quenched by the damp bone-chilling night mists rising from the Fens."[4]

In the Newnham library, along with the various scientific treatises that Cecilia was devouring, she found a manual for how to operate the observatory's equatorial telescope. It was a twenty-two-page stapled booklet, laboriously produced on a typewriter by someone, lost to history, who cared that a student at Newnham might want to put her eye to a scratched and weathered eyepiece.

"Allow the clockwork to go for about 5 min. until it has got up to speed," the manual advised. And "the clock often required a push to set it going."[5]

It was most likely on a cold, clear Cambridge night that Cecilia borrowed the key from the desk, left the library through the back entrance, and walked carefully in the dark past the garden and out into the large grassy field. By the thick-walled standards of gothic Cambridge, the Newnham Observatory was laughable. Its observatory dome was actually a pointed copper teepee set atop what resembled a little white-washed wooden barn surrounded by a bed of irises.[6]

Observatories are places where accidents happen. Astronomers work in the dark—they have been known to fall absent-mindedly into a telescope's well, or, carried away with observing, find their eyelid stuck to a frozen brass eyepiece. But on frigid moonlit Cambridge nights, the observatory's pale walls glowed ghostly white in the soft light of the distant library's reading lamps. It was a magical place.

Cecilia climbed the observatory's two steps, unlocked the door, and peered in. She stepped inside, ducking slightly—the building was apparently designed to admit only relatively short astronomers. In the dark, the telescope and the canvas-and-wood "observing chair" stood at the ready. That's when she made her first discovery: repairs were in order. The clock—required to slowly and smoothly drive the apparatus at the same speed that the earth rotates, but in the opposite direction, so that an object stays in the same place—needed more than "a push to set it going." It was, literally, frozen in time. She would need help.

She got it, in the form of the first Cambridge man she came in real contact with. His name was Leslie John Comrie, but everyone called him L. J. He was a character.

Newnham Observatory

L. J. Comrie was seven years older than Cecilia. He was born in 1893 in Pukekohe, Auckland, a bush-covered outpost of small farms on the northern tip of New Zealand. From his earliest years, L. J. had more energy than the other kids. He loved guns—he founded the Auckland University College Rifle Club—and sports—he was secretary of the college's tennis club. Although he earned both an undergraduate and a master's degree in chemistry, his passion was scouring the New Zealand night sky with a six-inch telescope on the roof of the Old Parliament Building.

As World War I raged, L. J. couldn't wait to get out of New Zealand and join the fight. Partly deaf since childhood, he tried over and over to enter the military. Finally, in April 1918, he "got in by a mixture of good luck and guile." The following fall, he joined the 1st Battalion of the 3rd New Zealand (Rifle) Brigade in France. He was with the new outfit

L. J. Comrie

for just a few days when an incoming artillery shell exploded close by. He lost a leg. It was, ironically, a British shell.[7]

Recovering in England, L. J. was not one to lie quietly in a hospital bed. He used the time to study at University College in London, where his interest turned from chemistry to mathematics. In 1919, Cecilia's first year, L. J. won a scholarship to St. John's College in Cambridge. He received Cambridge's prestigious Isaac Newton Scholarship the next year, and earned a PhD in 1923 for his analysis of planets in our solar system passing in front of stars, producing stellar occultations.

He was also not a man to suffer fools, or anyone else for that matter, who "did not attain his own high standards. He did not make allowances for the frailties of others, and was far from tactful in pointing them out."[8] Within the small tight-knit Cambridge astronomy community, he was a demonstrably determined man—partially deaf and one-legged, he still played a mean game of tennis—which may explain his being drawn to a demonstrably determined young woman. When Cecilia asked for help, he was quick to respond. He instructed her, as Cecilia recounted to an interviewer years later, on "what to do and how to clean out this and that, how to oil the clock and get it to run."[9]

With L. J. beside her, she opened the observatory's clock, removed a chrysalis from the works, got it moving again, and set about exploring the dark southern sky.[10] It was not easy working on the clock or focusing the telescope—they were bundled from head to toe (observatories were unheated because temperature differences between the air outside and inside would cause telescopic images to be fuzzy). Cecilia didn't mind; Newnham dorm rooms were drafty in winter, and she was always layered up. In truth, she liked the conditions. There is nothing like a cold dark night to focus the mind—no distraction from lights, no torpor from the comfort of warmth. "I discovered the beauties of the planets," she recalled later, describing that first of many nights. "Who can ever forget his first sight of the moons of Jupiter and the rings of Saturn?"[11]

She owned the place; it was her den and she was the wizard. As Betty Leaf described it, Cecilia was "now to be found on starry nights among the moths and dust of Newnham Observatory, showing the wonders of the Heavens to all who come."[12] She organized public viewing nights. She moved on from planets to variable stars, recording their changes from night to night. She bought a notebook and installed it next to the small measuring machine "with a notice that anyone who observed with the telescope must make a record in the book, and sign and date the entry."[13]

L. J. was a pioneer in the emerging discipline of computational astronomy. He had succeeded in predicting the eclipse of one of Saturn's moons by the shadow of another—a remarkable feat, in Cecilia's mind. She was in awe of him, and happily accepted his offer to give her lessons in computing. With his help, she began to assemble her own personal library of mathematical tables. With his recommendation, she joined the computing section of the British Astronomical Association in London.

It was not easy learning from Comrie. Stickler Searle at the Cavendish Lab was an amateur compared to him. "For Comrie, 'accuracy' had a specific and clear meaning, which could most often be acquired by the planning of elegant methods of work."[14] Under his demanding watch, if it wasn't right, it could only be wrong. Cecilia soon "found to my chagrin that I was an inaccurate computer and learned, painfully, to check my calculations—an invaluable training that has stood me in good stead."[15]

They were two extraordinarily driven people, Cecilia and L. J., and their conversations must have been focused and intense. Like Ivy Pendlebury at St. Paul's, like William Bateson and Agnes Arber in botany, even like cranky old Searle in the lab, drill-sergeant Comrie was exercising her mind, honing her analytical skills, toughening her to deal with hurdles both current and to come.

Cecilia's relationship with L. J. was nothing more than just respectful friendship. But for Cecilia, shy and subject to the university's strict rules, L. J.'s company would surely have been welcome. At Cambridge in the 1920s, the sexes were kept as separate as possible. Newnham women were inexperienced when it came to men because the college created an atmosphere in which there wasn't much opportunity to mix. One woman a few years older than Cecilia recalled that a vice principal once wondered aloud why girls needed to go to dances. "My sister mildly suggested that they might learn something about life." The vice principal's reply: "But what could they learn about life at a dance . . . that they couldn't learn just as well from a good book?"[16]

In Cecilia's first year, a group of Newnhamites staged a performance of *Belinda*, A. A. Milne's comedy about a middle-aged woman who alternates flirting with a dull statistician and a young poet who sports floppy hair and writes terrible verse. Because women had to take the men's roles, remembered one of the students, Principal B. A. Clough "decreed that only brothers or fiancés could attend because it was not suitable for any other men to see women in men's clothes!"[17]

All men had to be out of the college by half past six each evening. Clearly, though, some of the men didn't meet the deadline. In the minutes of a meeting of the Joint Committee of Staff and Students, it was recorded that Principal Clough wished to discourage "the growing practice of using Ground Floor windows as entrances and exits. She pointed out that the practice badly damaged the flower-beds below the windows."[18]

And if a woman did manage to scramble out a ground floor window, with or without a man, evading the bulldogs and getting back in before the breakfast roll-call was every bit as challenging. The simplest and most ingenious way, known by freshers and upperclass alike, had been devised by Newnhamites early on: "I've heard students of a later day boast of secret and complicated ways of circumventing regulations and slipping in late at night, but we had a simple plan—we found we could lie down flat and roll under one of the iron gates."[19]

For sure, there was push-back against the rigid traditions governing the relationship of men and women—gentle at first, but growing, especially among the postwar generation. It prompted the college to attempt to codify the unwritten rules. During Cecilia's second year, printed rules were handed out. For example: "The Vice-Principal must be informed if a man is to be entertained in the music room, and on no account must he eat or drink anything unless a chaperon be present." Lest there be any confusion: "No man is a competent chaperon but a certified cab-driver, father or brother." And finally: "Students may not be alone with a man either in a canoe or in a room, unless the man is either a fiancé or a brother."[20] The connection between canoe and room was left unstated.

So many rules, so many regulations—the only place of respite was the River Cam, flowing gently past the stone cliffs of the medieval buildings. It was a storybook scene—punts passing lazily under the bridges linking the various colleges as "a portable gramophone in the bows played Bach over the glassy stretches."[21] On the Cam, the formal campus rules gave

Punting on the River Cam

way to carefree leisure time. "Games, bicycle rides and the river took up much of our time, apart from work, and . . . woe betide any unlucky sinner who stayed late on the river, and appeared unchanged and dishevelled and had to engage in polite conversation with the dons [fellows of the college], conscious of burning cheeks and muddy shoes!"[22]

Comrie was almost a decade older than Cecilia, and an instructor, so they probably would have gotten away with being alone together on the river. There is no historical reference, but it is hard to imagine that they didn't now and then continue their computational conversations in a more relaxed setting, following the lead of the rest of the Cambridge community. "Everyone seemed to go on the river a tremendous lot," wrote a Newnham woman just a year older than Cecilia. "At week-ends the river between Silver Street and Grantchester was absolutely alive with punts, largely being manoeuvred by young men who didn't know much about punting, so there were frequent collisions and shrieks and people falling overboard."[23]

Could life in this melting pot of ideas dare to be carefree at times? Might even L. J. and Cecilia have taken a break from calculating nature to just experience nature's flow? As another Newnhamite described those days and that time: "The current carried us down between banks fragrant with hawthorn and golden kingcups. The world seemed new-created and, for half an hour, life, which we took so seriously, was fun."[24]

10

It was rarely quiet and peaceful at the University of Cambridge in the early twentieth century. So many sounds. Some were comforting: "the hours, and even quarters, chimed by innumerable clocks; the morning and evening ringing of the curfew from Great St. Mary's Church . . . the clop-clop on stone setts of the feet of the horses which drew the trams." Some were not so pleasant, as in 1918, the year before Cecilia arrived: students "standing breathless in the still sunlight of early morning to catch in the far distance the sound of the guns in France."[1]

For Cecilia, the sound that had become familiar now was the hissing of her bicycle tires on the pavement of Madingley Road. Astronomy had a firm hold on her, and one of the centers of astronomical research was only a fifteen-minute ride away. Today, when it is so easy to travel over great distances, observatories are built on isolated mountain tops under clear skies, and people come to them. In the nineteenth century, though, they were built where the people were, without regard for viewing conditions. Thus the construction of the Cambridge Observatory on the edge of town.

On the first night she visited the Cambridge Observatory, Cecilia would have pulled her bicycle from the rack just inside the Newnham Gate and set out. Down Sidgwick Avenue, a right turn onto Grange Road, then a left turn onto Madingley Road and on up the hill to the observatory's drive, the rough cobblestones rattling the fenders and the handlebar basket. So dark, so cold, but she was safe at least—this was Cambridge, and no one paid much attention to astronomers on bikes at night.

Whereas most other scientists do their work during the day, the astronomer's time for discovery is after dark. So it was at dusk that Cecilia turned into the observatory's grand entrance. Hedges lined either side of the road, opening at the end on to a manicured lawn and circular driveway. The observatory's main building was not Cambridge medieval, but it was no less imposing. The turquoise copper dome was supported by a stone portico, and a little red front door could be seen behind four grand Greek columns.

Initial funding for the main observatory building was raised by the Cambridge University Senate in 1820. A gravel pit on Madingley Road with a 360-degree view was selected, but the original equatorial telescope was not put in place under the dome until 1832. The head of the observatory was to be known as the Plumian professor, and the trustees of the professorship were intent on getting their money's worth—the live-in professor/astronomer's official duties were "to observe all night and to calculate all day."[2]

With six and a half acres of land, there was plenty of room to expand. Like a proud parent, the main observatory spawned a collection of domes scattered around the property, each atop its own unheated shelter, underwritten by UK merchants of great wealth and celestial curiosity. There was the Newall dome southwest of the main building, which housed a 25-inch-aperture telescope—in 1870, the largest telescope in the world—donated by the Scottish engineer Robert Sterling Newall. And just across a walkway was the Northumberland Telescope, a 12-inch equatorial refractor beneath a dome "mounted on cannon balls and revolving with great facility: 'a lady can turn it well without any machinery.'"[3]

Cambridge Observatory

Cecilia's destination that evening was yet another separately housed dome, just to the south of the main building. It was a telescope considerably larger than the one at the Newnham Observatory, one that was able to probe much deeper into the night sky. It too came with its own history.

Richard Sheepshanks was an early nineteenth-century trust fund baby. He graduated from Trinity College in 1816, qualified as both a barrister and an Anglican priest. He was saved from having to take a job in either bar or vestry, however, by inheriting the fortune left to him by his father, a wealthy Leeds textile manufacturer. Neither observer nor theoretician, Richard nonetheless adored astronomy, amassing an extensive collection of telescopes and serving as editor of the *Monthly Notices of the Royal Astronomical Society*.

He also apparently adored an Irish dancer by whom he had six children. When he died in 1855, he had no legitimate offspring. Accordingly, what was left of his inherited fortune passed to his sister, Anne Sheepshanks.

Anne then gave £10,000 to Cambridge with the express purpose of preserving Richard's name in astronomy.

The university used the funds to purchase, and name, the Sheepshanks Telescope, installed on the grounds of the Cambridge Observatory in 1898 under the directorship of the astronomer Robert S. Ball. It was meant to be used for astronomical photography: starlight would enter the lens, be directed down a long tube, and then reflected up to a camera inside the building. Although the telescope "had a fairly successful career," one astronomer sardonically remarked that the instrument "was of an unusual design which combined in a unique way the principal disadvantages of the refracting and reflecting forms of telescope."[4]

It was a regularly scheduled public night at the Sheepshanks Telescope. Cecilia, ducking her head and squeezing her tall frame through the hatch-like door, joined the crowd of the curious gathered around the 12.5-inch refractor. The observatory staff warned the visitors to "avoid the measuring machine with care." It was, as Cecilia recalled, "advice that I did not follow later on!"[5]

The kindly if somewhat brusque second assistant, Henry Ernest Green, made a show of prepping the equipment. Hand-over-hand, he worked a rope and pulley system that opened, with a deep rumbling, the south-facing observation slot—a curtain parting to reveal a starlit stage. He then yanked up the heavy iron falling weights of the timing mechanism—a far larger and more heavy-duty version of the one inside the Newnham Observatory. As gravity pulled on the weights, the clock drive smoothly turned the massive telescope, keeping it precisely aligned with the target. Henry then focused on an intriguing double star. As Cecilia looked through the eyepiece, he made sure she noted that the two parts were different colors.

"How can that be," she asked, "if they are of the same age?"

In his obituary notice years later, Henry was described as "quiet and modest, conscientious in everything he did, keenly interested in observatory affairs."[6] An astronomer, however, he was not. He was at a loss for an answer. To be fair, this was public visitor night; appreciative looks

Sheepshanks Telescope

of wonder were the norm, not difficult questions. As Cecilia pressed him, Henry finally threw up his hands in despair. "I will leave you in charge," he said to her, and quickly fled the scene.

And Cecilia did indeed take charge. Henry had turned the instrument to focus on the Andromeda Galaxy, the nearest large galaxy to our Milky Way, 2.5 million light-years from earth. Cecilia launched into telling the visitors everything she knew about Andromeda. "Heaven forgive my presumptuousness!" she wrote later about the incident.

Henry meanwhile raced down the stairs, across the dark grounds, through the little red door, and into the warm, softly lit study of the observatory's live-in professor. "Sir," he said, "there's a woman out there asking questions." He needed help.

By this time, Cecilia was holding up a young girl so that she could see through the eyepiece, and telling her what to look for. She heard a sound and turned around to find the observatory's ninth Plumian professor, Arthur Eddington, standing quietly behind her, listening. Cecilia may have been socially shy, but she knew when to seize a moment. She came right out and told Eddington that she would like to be an astronomer. The professor's laconic response: "I can see no *insuperable* objection."

Cecilia and Eddington then engaged in their own quiet conversation. She asked him what books she should read. He ticked off several. She told him that she had read them all already. It is likely that at this point Eddington began to realize he had more than just an interested student at his side. He referred her to the Royal Astronomical Society's *Monthly Notices* and the *Astrophysical Journal*. He told her that they were available in the observatory's library and that she was welcome to use it. Cecilia, true to form, would refer to another astronomer—William Herschel, the man who discovered Uranus—when describing how Eddington's offer affected her: "To paraphrase Herschel's epitaph, he had opened the doors of the heavens to me."

Eddington surely did not know at the time the significance of his simple, gracious gesture; nor did Cecilia, for that matter. But what she was able to do now was to begin to connect astronomy with physics. Even though she was not officially reading astronomy, she began to attend Eddington's lectures: "Relativity," "The Determination of Orbits," "The Reduction of Observations." Under Eddington's direction, the students would use logarithms to compute the orbits of comets.

Hours of calculating were followed by tea at the observatory—at the invitation of Eddington's mother and sister. Cecilia would join in with two or three other students. It was during these late-afternoon sessions, in front of the fireplace in the professor's study, that she got to know more about "the greatest intellect I have ever had the privilege to meet."[7] She learned that Eddington's favorite composer was a German—Engelbert Humperdinck—and that he leaned toward the music of the Scottish vaudeville singer Harry Lauder, for a time the highest paid performer in the world with his popular love song *Roamin' in the Gloamin'*.[8]

Cecilia described Eddington as "a very quiet man." His conversation "was punctuated by long silences. He never replied immediately to a question; he pondered it, and after a long (but not uncomfortable) interval would respond with a complete and rounded answer."[9]

Just like the Cavendish Lab, the Cambridge Observatory was home to a group of eccentric characters. W. M. Smart was ever present, lecturing on a range of astronomy topics from celestial mechanics to lunar theory. Smart had won the 1914 Tyson Medal for proficiency in astronomy and mathematics at Trinity College, but his lectures were often intellectual thickets, and he had little patience for students who couldn't keep up. As Cecilia put it, "He was not one to temper the wind to the shorn lamb."[10]

Frederick John Marrian Stratton, F. J. M., as he was known, was a decorated World War I colonel in the British Army and a lecturer at Cambridge on the emerging subject of astrophysics. He was one of the first professors to introduce Cecilia to the physical and chemical wonders of the stars. When Cecilia told him the same thing she had told Eddington—that she wished to become an astronomer—F. J. M. was not encouraging: "You can never hope to be anything but an amateur," he remarked.

Stratton was also fascinated by psychic phenomena. In later years, he would serve as president of the Society for Psychical Research, a group that specialized in the study of clairvoyance and the paranormal. Had the society been aware of F. J. M.'s poor predictive powers, however, it might have chosen another leader.

And then there was E. A. Milne. One afternoon Cecilia biked to the observatory grounds with, as usual, a question in mind. She "found a young man, his fair hair tumbling over his eyes, sitting astride the roof of one of the buildings, repairing it."

"Can I help you?" he called down to her.

Cecilia stood beside her bicycle. "I have come to ask," she shouted up to him, "why the Stark Effect is not observed in stellar spectra?"

The Stark Effect is a phenomenon in which spectral lines of atoms are broadened or split when passed through an electric field. The man on the roof could have been just a handyman, but in fact he was E. A. Milne, assistant director of the Solar Physics Observatory. Milne had won a mathematics scholarship to Trinity College in 1914 by gaining the largest number of marks that had ever been awarded for that particular grant. Unable to enlist because of poor eyesight, he had completed his military service by studying the physical behavior of artillery shells, hanging over the wings of aircraft in flight in order to take readings of temperature and pressure.

Milne came down off the roof and introduced himself. He could not have helped being captivated by a young woman on a bicycle asking the most penetrating questions. Like Henry Green, he did not know the answer to her question. Unlike Green, however, he didn't panic. Instead, he began counseling her, tutoring her in the principles of the emerging modern field of astrophysics. And, like L. J. Comrie, Milne would become to Cecilia, "a good friend and a great inspiration," their relationship extending well beyond the days of Cambridge.

Cecilia was meeting, and learning from, the very people who were formulating great theories and making major discoveries in astronomy, chemistry, and atomic physics. But now, in her remaining time at Cambridge, "the time for learning from others was not enough. I wanted to explore the frontiers for myself. I went to Eddington and asked him to introduce me to research."[11]

Eddington happily obliged. He gave her a problem to solve: integrate the temperature and density of a typical star, starting with the conditions at the center and then working outward, layer by layer. It sounded straightforward enough; but, as she would find out, Eddington was teaching her more than just the basics of stellar temperature and density. The problem haunted Cecilia; she worked on it, alone, day and night. She even dreamed of being at the center of the large bright star Betelgeuse, where the solution was obvious, until the light of day arrived. Perhaps the star's rotation should be factored in—but that simply produced more problems.

E. A. Milne

Exasperated, her incomplete solution in hand, she finally asked Eddington for help. He admitted that he had "been trying to solve that problem for years." It was tough love, but he was teaching her to realize that research could be enormously frustrating and that some problems were intractable.

Not all was in vain. Cecilia had earned the right to tackle a second problem: measuring the proper motion of stars located near NGC 1960, a cluster of stars well within the range of her familiar telescope. Eddington no doubt knew it was doable if she were resourceful enough. She spent "many happy hours (when I should have been in the advanced physics laboratory) sitting at the measuring machine in the housing of the Sheepshanks telescope."[12]

After she had made all the measurements, W. M. Smart, her advisor, directed her to the next step: fit the data points using the method of least squares. Cecilia was too proud to admit that she did not know the least squares method. She took a train to London, went straight to the reading room of the British Museum, and requested the works of Carl Friedrich Gauss, who had developed the method. As she remembered, "my heart sank when I was presented with five huge volumes, all in German."[13]

It was an obstacle, but a conquerable one. Cecilia was in her element—the hunt—and little else mattered. "She is most completely happy," her friend Betty Leaf had observed, "when some 'beautiful' mathematical theory of the Universe makes her forget the minor disturbances of everyday life."[14]

She studied the books, figured out the method, and wrote up her results. Eddington read her paper and pronounced it "very nice." She submitted it to the Royal Astronomical Society and "had the unspeakable joy of seeing my first paper in print."[15]

She had progressed—from learning from others, to learning by doing. She was also learning from herself, every obstacle overcome being a "valuable lesson." Thinking back on this period later, Cecilia mused that there are three goals that drive a person to want to become a scientist: fame, position, understanding. The first two could be obtained in any number of ways. Understanding, however, was "the only one of the three goals that continues to reward the pursuer." But understanding required what she called "intellectual integrity." A person with integrity would not be afraid to admit ignorance, and "an admission of ignorance may well be a step to a new discovery."[16]

She was also learning about things far beyond science. Pursuing the kind of study she wanted to do certainly would require intellectual integrity, but it also would require perseverance. She had to believe that pressing on would be rewarded, and that she would eventually uncover truths about the universe. For that kind of confidence, Cecilia turned not to science, but to poetry. When it got dark, she would recall the words of Wordsworth: "Knowing that Nature never did betray / The heart that loved her."[17]

11

The moment they stepped off the train at Cambridge station, the women students of Newnham College entered not just an academic world, but a social one as well. As interested as they might have been in flapper fashion and the new style for bobbed hair, they had to abide by the college regulations. One student a few years younger than Cecilia remembered: "To go about without stockings was unthinkable, even in summer in the garden. Two of my year took off their stockings once before going for an exercise run along Grange Road after dinner, and were sternly rebuked."[1]

The remainder of a woman's body was to be disguised beneath clothes at all times. "Dresses were loosely tubular, with a belt somewhere around the hips." The dons—only a few years older than the undergrads they watched over—went a step further, many of them adopting "a manly grey tailored suit with stiff collar and tie."[2]

The severe look for women could be traced back to a generation earlier. Even in Cecilia's time, but certainly in the decade before, women who had the temerity to enroll at Cambridge were looked upon by the male community more as visitors than as students. Trying to blend in, "we affected a mannish style in dress," remembered a fresher from 1907, "coat and skirt, blouse with a stiff collar and knotted tie, although our skirts nearly touched the ground." Only on the wide open hockey field did the prudish give way to the merely prim: "Skirts worn for games might be a few inches off the ground," remembered a woman about her college days around 1900. And even then, "one of our team had to write home for permission to leave off her flannel petticoat when playing hockey."[3]

If fashion was a way of speaking, defying certain fashion norms was a way of speaking out. Hairstyles provided one way to do so. A decade earlier, hair had been "done up with a hundred hair-pins in buns at the back of the head or on top." By the time Cecilia's class arrived, scissors had come out: women wore their hair in "short thick bobs. This fashion swept through the College in our first year; and the little hairdresser in Regent Street must have made a fortune shearing off our manes."[4]

Newnham students were starting to rebel against being "mouselike," against blending in. There were small but real expressions of individuality, and they did not go unnoticed. Just five years after Cecilia graduated, Virginia Woolf was invited to address a crowd of eager undergrads at Newnham. The talk would be featured in her book *A Room of One's Own*, published a year later. The post-address gathering, appropriately enough, was held in a dorm room.

The young woman who lived in that particular room had to have been relatively upper class; Woolf, who made her name by noticing, took notice. As the resident of the room later recalled, "I think I had expected some profound, philosophic remarks, even after prunes and custard; but fixing me with that wonderful gaze, at once luminous and penetrating, what she actually said was, 'I'd no idea the young ladies of Newnham were so beautifully dressed.'"[5]

If Cecilia had been in that room, her outfit would certainly not have elicited such a comment. Her various scholarships did not leave room for much discretionary spending on fine clothes. She would not have frequented the high-end stores of Magdalene Street or King Street; she

Cecilia Payne, 1922

would more likely have been found at a "little arty-crafty shop in King's Parade [which] did a brisk trade in rather lumpy hand-woven linen jumpers and skirts with patterned edges and fringes."[6]

Fashion was low on her list of priorities. She suspected that her disinterest affected how she was looked upon in the lecture halls and laboratories, where "women were treated as second-class students." "It might have been different if I had been gay and attractive and had worn pretty clothes," she wrote later. "But I was dowdy and studious, comically serious and agonizingly shy."[7]

Dowdy, serious, studious, shy—qualities that were welcomed, even prized, at the Cambridge Observatory. But Cecilia knew that she could not investigate the makeup of stars, planets, and comets simply by reading astronomical journals and gazing through telescopes. Heavenly bodies had physical properties, which meant that the key to understanding them was behind the door of the Cavendish Lab, not the Cambridge Observatory.

When she parked her bike, took off her hat, and pushed through the lab's thick entrance door, she would step into a world that was bubbling over with ideas and experiments and egos. Her timing was perfect; she was able to learn about the new physics from the very people who practiced it. "The study of physics was pure delight," she wrote later. "The Cavendish was peopled with legendary figures."[8]

The head of the lab when Cecilia first ventured in was in fact one of those legendary figures: Joseph James Thomson. Thomson, who was always known as J. J., had prowled the lab's hallways for thirty-five years. "Matters have come to a pretty pass when they elect mere boys Professors," sniffed a senior member of the university back in 1884 when J. J., then only twenty-eight, was appointed the third head of the Cavendish Lab.[9]

He and the lab were made for each other, for J. J. was the ultimate measurer. At the time, the atom was still regarded as the smallest, indivisible building block of nature, but J. J. was beginning to think otherwise. Fusing chemistry with physics, he pictured the atom the way Lord Kelvin had described it in 1867—as a "smoke ring"—and then decided to see where that might lead him. In his Cavendish cubicle in 1897, he experimented with what were known as cathode rays. These were strange beams of light that appeared inside a glass vacuum tube when an electric current passed between a cathode at one end of the tube and an anode at the other end.

Thomson moved magnets and charged plates around the tube and experimented with different vacuum levels. He concluded that the cathode rays were composed of negatively charged particles that were much smaller than a single hydrogen atom—in fact, a thousand times smaller.

This finding shattered the centuries-held belief that atoms were the smallest building blocks of nature. Now physicists and chemists began

J. J. Thomson

to focus on *subatomic* particles. As brilliant as J. J. was as both theorist and experimenter, however, he was a terrible marketer. He named his newly found subatomic particles "corpuscles." Fortunately, a few years earlier another Cambridge-educated physicist, one George Johnstone Stoney, had proposed the word "electron" to describe "the fundamental unit of electricity."[10] J. J. won the Nobel Prize; Stoney's term stuck. And although J. J. could not have known it at the time, he, or at least the Cavendish Lab, should have taken out a patent. The cathode ray tube would become the foundation for an entire industry: television.

J. J. was also a classic example of the men's club scientist. Women, in J. J.'s opinion, simply did not have the intellectual capacity to be world-class physicists. "I have got one at my advanced lecture," he wrote in a letter to a family friend. "I am afraid she does not understand a word and my theory is that she is attending my lectures on the supposition that they are Divinity and she has not yet found out her mistake."[11]

When the head of an institution has forceful opinions, they filter down to subordinates. The laboratory assistants—the lab boys, as they were known—clearly sensed J. J.'s disdain for women. As if the study of science at Cambridge were not difficult enough, "the lab boys took a delight in leaving some essential bit of apparatus out of our lists so that we had to walk the whole length of the lab to the store to ask for it," recalled a Newnham student admitted into the lab in 1910. "An ordeal for some of us, especially as they appeared to be too busy to attend to us for several minutes while we waited at the door."[12]

But as important as research was, the Cavendish Lab, per Maxwell's original vision, was to be a center of learning. The physicists in residence might win Nobels for their experiments, but all of them, including the director, were expected—in fact, required—to teach.

J. J.'s teaching style was as singular as his lab work. He "shaved badly" and wore "his hair rather long," according to colleagues. And he steadfastly adhered to the traditional view that women students should be separated, at all times, from the men. Women had to sit in the front rows of lecture halls and on separate benches in the labs. One upper-class woman described the situation in Cecilia's first year: as women students "walked down the steps of the big lecture theatres to their places in the front row, every man behind them clumped and stamped in time with each of their steps."[13]

It was as if enduring this crowd behavior was the women's price of admission to join the club. Moreover, it was just after the war. Cambridge—blackouts suspended, church bells pealing—was filled with returning soldiers, and women were resented. As the same student recalled: "I remember on one occasion leaving my dissecting scissors behind when I came to work in the Anatomy Lab. Though surrounded by men who all had extra pairs of scissors, I felt it necessary to bicycle back to Newnham to get my own!"[14]

Sir Ernest Rutherford

For over three decades, J. J. was a masterful head of the lab; seven Nobel Prizes were awarded to Cavendish professors during his reign. But toward the end of Cecilia's first year, in 1919, his name and "fossil" had begun to be heard together in conversations. With the handwriting on the laboratory wall, J. J. finally agreed to step down to become master of Trinity College, paving the way for the Cavendish Lab to appoint a new director—an aggressive, blunt, eccentric, tough-minded New Zealander: Ernest Rutherford.

Rutherford was not new to the Cavendish Lab. Twenty-five years earlier, he had been the lab's first "alien"—a graduate student without a Cambridge degree who was allowed to do research at the university.

Rutherford actively sought out physics problems to solve. The first one he latched on to, in 1895, was the detection of electromagnetic waves. Like an astronomer late at night on a bicycle, a quirky young man in his twenties did not draw much attention as he set up a transmitter on top of the tower of St. John's College chapel and then walked backward down Portugal Street with his detector. He briefly held the record for the distance at which radio waves could be detected: half a mile.

Rutherford won the 1908 Nobel Prize in Chemistry for his work in radioactivity. His reputation in physics began to build when he coined the terms alpha, beta, and gamma to describe various forms of radioactive radiation. But fame burst into the open with the results of his "gold foil experiment." When he fired alpha particles (helium nuclei produced by radioactive decay) at a thin piece of gold, some of the particles bounced straight back, apparently having struck something massive. The plum pudding model was dead.

"It was quite the most incredible event that has ever happened to me in my life," he recalled later. "It was almost as incredible as if you fired a 15-inch shell at a piece of tissue paper and it came back and hit you."[15] Rutherford had discovered that the mass of an atom was concentrated in its nucleus. It was the birth of nuclear physics.

Rutherford became the head of the Cavendish Lab in 1919, Cecilia's first year at Cambridge, and he would reign over the lab until the day he died, almost two decades later, in 1937. So much about how the physical world works would be discovered in that lab over that time span; it has been described as the final act in the story of nuclear physics, with Rutherford being to the atom what Darwin was to evolution. He became a magnet, drawing scientific minds and worldly glory to the lab. The people who came to work with him were called Rutherford's Boys, and they were indeed all boys.

The lab was busy and it was crowded. Rutherford knew collaboration produced results, and he encouraged teamwork. Research groups formed, led by strong individuals, and they all worked alongside each other. Francis Aston investigated isotopes and made precise measurements of atomic weights. Mark Oliphant and John Cockcroft explored the acceleration of charged particles. Hans Geiger and Walther Müller worked on detecting radiation. James Chadwick discovered the neutron, making it possible to understand at last what the atomic nucleus was made of.

Added to the research mix were thirty or so undergraduate students each year. Rutherford would walk around the hive of activity, sliding onto a stool and peppering with questions the shy, intensely focused young physicists at their work benches, their hands occupied with wires and meters and gauges. He set a formal tone, with his three-piece suit, thin tie beneath a turned-up collar, thick groomed moustache, and, on cold days, a black topcoat and black bowler hat. Even on the rare occasions when Rutherford would change out of his suit to work on his own experiments, decorum came first. He was known to hang back at afternoon tea parties if he was wearing his lab clothes and had not shaved.[16]

A lab filled with competitive experimenters had the potential to get unruly, so Rutherford imposed on everyone a get-it-done sense of discipline. Unlike in astronomy, the day was for work, not the night. "Punctually at six o'clock each evening, the senior laboratory assistant would tour the laboratories announcing to all that it was time, gentlemen, to close." It was Rutherford's dictum that "if one hadn't accomplished what one wished by six o'clock, it was unlikely that one would do so thereafter."[17]

Powerful insights into the physical world came from devising instruments that were straightforward in construction and easy to calibrate. "Simple, unpretentious appearances, but striking inferences: these were the Cavendish trademarks."[18] The result was that Rutherford's Cavendish Lab was a scene of scientific frenzy: experiments, observations, and discoveries the likes of which the world had never seen before. Five Nobel Prizes were awarded to Cavendish physicists during Rutherford's reign.

"The Professor," however, was still exactly that, which meant Rutherford was expected to teach classes, just like any other professor. He lectured on topics such as "the constitution of matter" at noon on Mondays, Wednesdays, and Fridays in the Maxwell Lecture Theater, a cavernous room thirty-eight feet long, thirty-five feet wide, and twenty-eight feet high.

As the noon hour approached, the theater's wooden plank floor would creak as Cecilia and thirty other students filed into the room. At the stroke of twelve, Rutherford would stride in, pulling a few notes from the inside pocket of his suit jacket. With half his mind thinking about the day's topic and the other half focusing on one experiment or another, he would frequently lose track of what he was saying. His legendary impatience would then flare as he scowled at his audience: "You sit there like a lot of numbskulls, and not one of you can tell me where I've gone wrong!"[19]

"There was no doubt that we were listening to a great man relating an epic story," remembers one former student. "Rather like the story of some great scientific exploration as told by its leader."[20]

The Maxwell Theater featured an enormous blackboard at the front of the room. The listeners sat upright on hard wooden benches, using the back of the row in front as a narrow desk. If you were unlucky enough to find yourself in the front row, you took notes as best you could.

"The advanced course in physics began with Rutherford's lectures. I was the only woman student who attended them and the regulations required that women should sit by themselves in the front row," Cecilia recalled later. "At every lecture Rutherford would gaze at me pointedly, as I sat by myself under his very nose, and would begin in his stentorian voice: '*Ladies* and Gentlemen.'"[21]

Did he mean it to be derisive? Perhaps not; maybe he thought it was just good-natured fun, for Rutherford would later defend women's rights in higher education. But there is no question that the rest of the class saw it that way. As Cecilia described it, "All the boys regularly greeted this witticism with thunderous applause, stamping with their feet in the traditional manner, and at every lecture I wished I could sink into the earth."[22]

It could not have been easy—trying to master the principles of atomic physics, scribbling in a notebook held awkwardly in her lap, in the spotlight and under the gaze of the entire class. It is remarkable that she got through it. But Cecilia realized how fortunate she was to be there. For centuries, scientists had been trying to uncover the basic building blocks of nature; and now, at this particular point in history, it had all come together. Here she was, listening to long-sought answers as told by the very men who had cracked the code.

There is another reason why Cecilia persevered: she had a need to learn, to observe. She endured everything from laboratory slights to classroom derision because there was no choice. She was driven to understand, which meant that nothing in the way would stop her.

It was trial by fire with Rutherford. He was doing for her what Bateson had done before, though in a far more sustained and intense way: he was training her mind. Rutherford had said over and over that theories were basically opinions; what really mattered were facts. But in addition to appreciating the facts—the properties of x-rays, the phenomenon of radioactive decay, the creation of isotopes—Cecilia was absorbing Rutherford's get-it-done attitude and his relentlessly competitive drive to understand.

Like most of those listening and watching in the theater, Cecilia was impressed by how Rutherford could take a complicated subject, atomic physics, and make it seem simple. All matter was composed of the same fundamental building blocks, and the simplest element of all was hydrogen: one proton, one electron. Bohr had used this model to explain hydrogen wavelength patterns in terms of his quantum theory. In the choice between complex and simple, Mother Nature would choose simple.

Cecilia may have noticed a common thread running among all these historic breakthroughs: the discoverer had managed to fuse physics with another discipline. Applying physics had enabled these men to view long-held scientific problems from a new, completely different angle. Rutherford had fused physics with chemistry to discover the proton; Bohr had fused physics with the quantum concept to explain how the atom behaves;

Einstein had combined physics with mathematics to derive his formula for the relationship between energy and mass.

What if she could do the same? A star was not something one could analyze under a microscope or observe inside a glass tube. How was a scientist to experiment on the unapproachable, on something light-years away? Could her growing knowledge of the building blocks of nature be used for learning something about the heavens? Could she fuse physics with astronomy and bring stars to earth?

12

"The thrill of acquiring knowledge" was one of the things at Cambridge that "really mattered," in the words of a classmate of Cecilia's at Newnham College—a place where young women hungry to learn felt comfortable in an otherwise male-dominated campus.[1]

But there was tension in the air; it had been building for years. All those little pushbacks against Victorian control—wearing shorts on the river, taking control of a motorcycle, cutting hair short, removing stockings before a run—individually they were not of much significance, but collectively they were growing into a force.

Pressure had been building on Blanche Athena Clough. "B. A." was a dignified woman who repeatedly turned down the job as principal of Newnham until her colleagues finally prevailed on her to accept it in 1920, Cecilia's second year. Two years into the job, her charges were getting restless. Cambridge women studied the same materials, took the same tests, learned the same material as Cambridge men. The sole difference was that women were not awarded degrees. While women at Oxford were granted degrees in 1920, not so women at Cambridge.

Pushbacks came to shoves at the start of Cecilia's third year. Two pro-
posals had been submitted to the university's Council of the Senate: (1)
admission of women to full membership in the university, but exclusion
from the Senate; (2) no admission, but women could be granted "titular
degrees." The second passed but the first was rejected, which meant that
women had to settle for being granted titles of degrees, but not the degrees themselves.[2]

Uproar! But not from the women. Cambridge men were furious that
women students had been so uppity as to seek equivalent recognition for
equivalent effort. Women had been given the right to use the libraries—
wasn't that enough? It was a charged atmosphere. Even though the
women had been defeated, when the results of the vote were announced
on the night of October 24, 1921, to a crowd of men, young and old alike,
outside the Senate House, "one of these elderly warriors so far forgot himself as to utter inflammatory words urging the young men to proceed to
Newnham College—with what precise object he did not specify."[3]

There had always been student demonstrations, but they were usually
more or less harmless—undergraduates being generally rowdy during
the annual Guy Fawkes bonfire night, or members of the Pavement Club
occupying King's Parade and diverting traffic, or student Egyptologists
taking over a market place for excavation purposes. Those inconve-
nienced were mostly the police, who, after one of these disruptions,
"were not amused and lost one or two helmets—these being precious tro-
phies in undergraduates' rooms."[4]

The "elderly warrior's" call to arms, however, was not an annual ritual.
Old black-and-white newsreel footage shows a huge throng of men,
some dressed as women, marching through Cambridge streets holding
banners:

LOOK AT US

DON'T WE

DESERVE DEGREES

NOW YOU HAVE

SEEN US! GIVE US

OUR DEGREES!![5]

One Newnham student described the scene: "Filled with explosive energy and devoid of any idea as to what to do with it, they surged up Newnham Walk in the direction of . . . the main entrance. Here unfortunately they came upon one of the long-handled, four-wheeled trolleys used by the porters for distributing coal. This they seized, and used as a battering-ram to smash down the Clough Memorial Gates."[6]

For Cecilia and the other Newnhamites, their college now "resembled a beleaguered fortress which we were forbidden to leave." As they gathered in dorm rooms overlooking the mob, they witnessed a remarkable scene: "Miss Clough standing in . . . the archway, facing the gates with the howling mob just outside." The ornate wrought iron gates were bent and broken, but they held. As for the mob, B. A. Clough—described as showing "great dignity and calm"—simply stared them down.[7]

In the hung-over stillness of the next day, the university officially apologized to Principal Clough. And in the days following, the iron gates were repaired. But that was all. It would take another twenty-six years for the University Senate to decree, in 1947, that Cambridge women be given, in person, the degrees they had earned.

In this somber air, Cecilia focused on finishing her college work. In one sense, her distinctly middle-class background was an advantage— she was expected to learn. "Young women from the most upper class backgrounds . . . often faced stronger opposition than women in other classes to going to university and becoming, as their families feared, unmarriageable 'bluestockings.'" In fact, she was more than expected to learn; she was required to. "Parents of women graduates of the 1920s . . . were preparing their daughters for the real possibility that they might not marry and would have to be self-supporting throughout their lives."[8]

The irony was probably not lost on Cecilia. She was middle class and smart, which meant she might not get married, which meant she had to get a good education, which meant she had to secure a job but without the degree as proof that she could do the work. It made for interesting late-night conversations: "We were discussing whether a Newnham student could be allowed to take an afternoon walk with an undergraduate," recalled a student a year ahead of Cecilia. Another student replied, "'Well . . . yes . . . if she is engaged to him.' To which a friend of mine . . . replied that it seemed a heavy price to pay for a walk."[9]

In fact, as another of Cecilia's classmates dared suggest, might there not be a higher calling? "Many of us had the sense that we were a special dedicated group who would not marry because we had more exciting things to do. There was even a sense that a girl who got engaged was almost letting the side down."[10]

In society's view at the time, however, there were not more exciting things to do. Virtually the only profession open to women who did not marry was that of schoolteacher. Cecilia considered the idea until she came across *The Compleat Schoolmarm*, a slim book that she and her classmates "passed from hand to hand with cynical chuckles."[11] It painted a grim picture of the limited life and small horizons of a schoolmistress. It was, and is, a tedious read.

Cecilia was swimming against the tide. Because women at the time had been conditioned to believe that being feminine required self-sacrifice, that a woman should always be quietly attentive, "it was not surprising that the majority of girls were . . . unambitious, and those who *were* ambitious often felt embarrassed or guilty about being so."[12]

Cecilia Payne had endured derision in the laboratory and the lecture hall, studied early quantum mechanics, worked right alongside Rutherford's Boys at the Cavendish Lab and Eddington at the Cambridge Observatory—all to become a schoolmarm? "I saw an abyss opening before my feet," she recalled. "My taste of the world of scientists had unfitted me for such a calling."[13]

L. J. Comrie pulled Cecilia back from the abyss. It must have been clear to him that Cecilia Payne was not cut out to be a schoolteacher. He had tutored her and witnessed her intense need to probe the secrets of the universe. He was absolutely certain she could be a brilliant astronomer. L. J. confirmed what Cecilia already suspected—that there was no chance for her to find work as a scientist in England. He told her that he was soon going to take a teaching position at a college called Swarthmore in the United States. He told her that if she wanted to be an astronomer, she, too, should go to America.

And how exactly was that going to happen? L. J. suggested that Cecilia come with him to London. He would bring her as his guest to a lecture to be given to the British Astronomical Association by Harlow Shapley, director of the Harvard College Observatory. He would introduce her. She agreed to go.[14]

It turned out to be a lecture Cecilia would not forget. Shapley was not eloquent, in Cecilia's opinion; she thought he lacked Eddington's polish. She was also taken aback by Shapley's offhand levity. He showed the audience a photograph of Messier 8—a giant interstellar cloud in the constellation Sagittarius—and pointed out the various dark patches within the nebula. "Some people might say that they are the fingerprints of God," Shapley intoned, then paused for effect before continuing. "But perhaps they are only the fingerprints of the careless devil who made the plate."[15]

But like Eddington, Shapley knew how to grab an audience. Even Cecilia, who was more familiar with the subject than most in the room, was captivated. "Here was a man who walked with the stars and spoke of them as familiar friends," she wrote later. "They were brought within reach; one could almost touch them." As she had done when Eddington spoke, the next day she wrote down everything Shapley said, word for word.[16]

And just as she also had done with Eddington, she came directly to the point. When L. J. introduced her to Shapley, she said, "I should like to come and work under you." A number of women had been employed as "computers" in the Harvard Observatory, and at the time Annie Jump Cannon held the position of curator of astronomical photographs. Shapley professed to be delighted at the prospect of a smart young woman from Cambridge working for him. "When Miss Cannon retires, you can succeed her," he answered.[17] It was an offhand response made in a room crowded with astronomers. Cecilia and L. J. departed quietly.

Back in Cambridge, Cecilia carried on, with things both great—physics—and small—cocoa. The cocoa party was a ritual dating back to Newnham's earliest days. Students would bring large mugs of milk back after dinner and then invite friends to their room for hot chocolate. The result was a forced fostering of friendships, an example of "the intimidating etiquette that encrusted almost every detail of student life." And the rules of etiquette were as binding as, well, a corset: "A condescending second-year could invite a gratified first-year to cocoa; but if an exalted third-year proffered a cup across the gulf between eighteen and twenty, it had to contain coffee and the invitation must be for lunch on Sunday."[18]

The nightly custom of imbibing powdered chocolate may have built lasting friendships, but there were also unintended consequences. One woman, a year older than Cecilia, noted: "At the end of our first term our families remarked on our buxom appearance, due to the rich midnight brews of sweet milky cocoa, boiled on the coal fires in our room."[19]

And then there was propping. "May I prop?"—as in, "Shall we call each other by our Christian names?" It was another Cambridge custom, with its own cocoa-style class system: "Any first-year might prop any other first-year but only a second-year or, in rare cases, a third-year might prop a fresher. The up-and-coming freshers then had a delightful time name-dropping, while the down-and-going freshers, drooping unpropped, suffered corresponding dejection."[20]

One of the classmates Cecilia propped was Ernest Rutherford's daughter, Eileen. Cecilia drew close to her, describing her as "a lovable, spontaneous girl, without an ounce of science in her makeup." Eileen invited Cecilia to her home for tea, but later she told Cecilia that her father had remarked, "She isn't interested in *you*, my dear; she's interested in *me*." It angered Cecilia, although to herself she admitted candidly that "there might have been a grain of truth in this remark." Nevertheless, she never entered their house again.[21]

At the lab, Rutherford loomed as large as ever. The physics—perceived, rightfully, by Cecilia to be more and more important to the study of astronomy—was difficult enough without what she also perceived as Rutherford's "scorn for women." And when she made mistakes, Newnham was no help; the college could not provide a tutor in advanced physics. She bounced from one bored young physicist to another.

The student advisor for the advanced physics laboratory was thirty-three-year-old Henry Thirkill. A lifelong Mason who favored dark suits, white shirts, black ties, and black shoes, Thirks, as he was known in the physics fraternity, had little regard for the young woman assigned to him. He once told a colleague that he thought Cecilia was "slow." That assuredly was not the case. "Ignorant and uncouth I might be," Cecilia thought to herself on hearing about the remark, "but not *slow!*"[22]

Years earlier, Betty Leaf had written: "The most surprising thing about Cecilia is her many-sidedness."[23] Being politic, however, was not one of those sides. After the "slow" comment, she "decided to pay no more attention to anything Henry Thirkill said." Not good. Thirks happened to be one of the questioners for her final physics exam, known as the Tripos. She thought later that he might have been responsible for placing her in the second class in the exam results. She passed, but "not too creditably," as she recalled later. But she also had to admit that perhaps Thirks was not entirely to blame. "When I remember how I neglected my studies for work at the Observatory, it is surprising that I secured even a Second Class in the Tripos."[24]

13

As Cecilia's days at Cambridge wound down, she was faced with the prospect of what to do upon graduating. Had she graduated a few years earlier, one option would have been to join the "Steamboat Ladies." From 1904 to 1907, another Trinity College, the one at the University of Dublin, had offered actual degrees to all women graduates of Cambridge who had received the title only. The fee, not including steamboat passage to Dublin, was £10. A generic "Dear Madam" letter in the principal's office explained the procedure and noted that "several former Newnham students have already been admitted to the degree."[1] That list included Cecilia's former headmistress from the St. Paul's Girls' School, Frances Ralph Gray. It was a circuitous route; but if you were a woman who wanted or needed parchment in hand at that time, that's what you did.

Cecilia was not able to follow that path. Moreover, she had to know that L. J. was right; there was no hope that she could find meaningful work in astronomy in England. The Harvard Observatory job was her only ticket to doing what she had spent years preparing for.

She needed to do something fast. It had been eight months since L. J. had introduced her to Harlow Shapley. She sat down on February 26, 1923, and composed a heartfelt letter to the Harvard Observatory director. It was two full pages, written in the careful penmanship of a studious twenty-two-year-old. It was charming. It was also direct. She reminded Shapley of their meeting and then immediately mentioned Eddington, "under whom I had been doing some work. . . . He advised me most strongly to come to Harvard if I could."[2]

She told Shapley that if she could secure a fellowship, she hoped to be able to work at Harvard for a year starting in September. She began with a plea for "a Research Fellowship of $650 which I understand is offered at Harvard University." She made clear her desire but also her financial limitations: "I am extremely anxious to come if it is possible, and prepared to undertake anything which would enable me to work at Harvard, as I have not any private means to do so."

She ended the letter by repeating her need for funding: "I am trying to arrange to come to work at Harvard at the advice of Professor Eddington. . . . My means are quite inadequate for this, and it is necessary that I should obtain a Fellowship or other grant, in order to be able to come." She included her CV: two more handwritten pages crammed with descriptions of academic qualifications, courses completed, languages learned.

She marshaled all her forces. She prevailed on Eddington to write a recommendation to Shapley, and he wrote a glowing one: "She has attained a wide knowledge of physical science including astronomy, and in addition possesses the valuable qualities of enthusiasm and energy in her work."[3]

She prevailed on L. J. Comrie. He made sure Shapley knew that he was nominating Cecilia to the Royal Astronomical Society, writing: "I know of no lady in England who is more likely to be successful at Harvard than Miss Payne." And in case Shapley had any reservations based on her gender, he added: "There is this to be said about her (between ourselves)—I believe she is the type of person who, given the opportunity, would devote her whole life to astronomy and that she would not want to run away after a few years training to get married."[4]

She also prevailed on the headmistress of St. Paul's. Frances Gray testified to her "whole-hearted devotion to research and [her] intelligent and well-thought-out methods of work."[5]

She even managed to enlist stickler Searle, who described Cecilia as "a thoroughly earnest student and very keen on her work." Searle was not as enthusiastic as Eddington or Comrie. He thought she would never become a true astronomer, and that she would not last long in America. But a year at Harvard, he wrote, "would make her valuable as a teacher or investigator in England."[6]

Shapley was impressed. The recommenders were luminaries of astronomy and academe, and Cecilia's four-page package radiated intelligence, drive, eagerness, competence. He was careful, though, with his reply. With a penny-pinching theme that would continue throughout the years to come, he promised only that "if you are successful in obtaining a fellowship I shall be glad if you will carry on your astronomical researches at the Harvard College Observatory next year under my direction."[7]

Cecilia responded immediately upon receiving Shapley's encouraging letter. She listed all the stipends and scholarships she was applying for. She also wrote that it was L. J. who had pointed out to her that Harvard had the Pickering Fellowship—worth $650—to offer, and that she had formally applied for it. And then, shy Cecilia Payne made sure Shapley knew that getting her would be a bargain. "If I could have such a Fellowship," she wrote, "it would be definitely possible for me to come and I should be very grateful indeed."[8]

In his reply, Shapley acknowledged that she was indeed a bargain. "Whether or not a Pickering Fellowship will be available for you for next year I cannot yet say," he wrote back. "In any case, if it is necessary, you may count on a stipend of at least $500 from here."[9]

Relieved but still wary, in her reply, typewritten now, Cecilia responded that Shapley's cash offer (meager as it was) "makes it appear most likely that I shall be able to come." She added that she would let him know as soon as possible "how my plans turn out."[10]

In truth, she probably would have accepted even if she only had Shapley's $500. With his offer in hand, though, she now "bent all my efforts to getting the support I needed, and collected enough money in Fellowships and grants to finance a year in the United States."[11]

She sealed the deal in a letter of June 23 to Shapley. She informed him that Newnham had offered her a Bathurst Studentship worth £120. "If I may have in addition the $500 (which you promised me if I should need it) this makes it possible for me to come to Harvard in September."[12]

What she didn't tell Shapley about was the pocket money she had obtained for new clothes. She had "made a shameless search of the College catalogue for a profitable prize," she related later. "There I found an offer of £50, a veritable fortune, for an essay on the Greek text of one of the Gospels." Back she went to London, to the British Museum, to the same reading room where she had studied Gauss. She compiled a digest of religious treatises and wrote a lengthy essay, chock-full of Greek footnotes. "As I had guessed correctly," she recalled, "the subject had attracted no other candidates, and I was able to enjoy my first earnings, which I spent on outfitting myself for my journey to the New World."[13]

With enough cash now in hand, she booked passage on RMS *Laconia*. The ship was a new Cunard liner; it had made its maiden voyage from Southampton to New York just the year before. (The *Laconia* would manage to survive a number of collisions with other vessels in the years to come, only to be sunk by torpedoes from a German submarine in 1942.)

There was no turning back now. "In the fall of 1923, I prepared to leave my native land, to sail to the west as more than one of my ancestors had done. The time for dreaming was over. I was about to enter the real world."[14]

It was time to spread the word. When she told Eddington, he smiled and quietly wished her well. When she told L. J., his response was both promotional and practical. Preparing to go to Swarthmore to teach, he was enthusiastic about the United States. He pointed out that "women had more opportunities for astronomical research in the United States than in England."[15] He also told her he would meet her ship in New York and make sure she got on the train to Boston.[16]

When she told E. A. Milne, his reaction was more philosophical than congratulatory or practical; it would prove prophetic as well. Milne was pursuing a new line of thought in the study of stars. It was known that certain patterns, or lines, of the color spectrum could be produced in a laboratory only at high temperatures or under immense pressure. He knew that similar lines were often found in the photographed spectra of stars. But why? What explained the relationship between the earthly phenomenon and the stellar one? Milne told Cecilia that if he were the one going to Harvard, he would use Harvard's data to experimentally verify what was known as Saha's equation.[17]

Meghnad Saha was a physicist in India who had developed an equation that related a star's temperature and pressure to the lines in its spectrum. He was attempting to link the spectrum of the sun and other stars to Bohr's description of the atom. But it was all still just theory; Saha could not demonstrate that his calculations were correct because he had no access to raw data. Cecilia, on the other hand, had been taught by Bohr, had learned experimental atomic physics at the knee of Rutherford, and was headed to a facility that had a huge trove of data in the form of hundreds of thousands of stellar spectra. Think about it, said Milne.

She did think about it. What Milne was describing was nothing less than the birth of astrophysics. One could not travel from earth to a star, take a sample, and analyze it. But a star's light could and did travel, all the way to earth, its journey's end a photograph on a glass plate. She knew that the starlight's spectrum could then be analyzed if the researcher had the right knowledge. So, might the unique pattern of a star's spectrum be the key to understanding what the star is made of?

It is hard to imagine what thoughts were occupying Cecilia's mind at this moment. She had scraped together the means to study in America. She would be able to do what she so desperately wanted, research in astronomy. But to do so required leaving the bee orchis of Wendover,

Cecilia on board the RMS *Laconia*, 1923

leaving the "brooding shadow" of London, leaving the "intellectual integrity" of Cambridge. Leaving her brother, Humfry, and her sister, Leonora. Leaving her mother, Emma.

On September 10, 1923, she packed her new clothes in a trunk. She placed her inheritance—her father's violin—in its case. She put on a long black overcoat and a wide-brimmed black hat. With the clothes trunk and the violin case in hand, she took a train to Southampton—not to board a steamship bound for Dublin, but to board an ocean liner bound for New York.

She did not let her sadness at leaving her family and homeland stop her. She had not had much contact with her mother and siblings in recent years. She was not like the few schoolteachers-to-be, much less the stay-at-home girls. Friends and family did not quite know what to make of her. She was not the typical English schoolgirl of the time, who "had learned from her childhood that worldly success was unfeminine, and even in areas where women *might* achieve success, it was unfeminine to *seek* to earn it."[18]

Cecilia fit no model, fit no neat little box. She *was* feminine, but she *was* going to seek success. And she *was* going to earn it. That is not to say she didn't miss her country and her family. She did. Especially her extended family. She put a snapshot of Betty Leaf in her wallet.

III

DISCOVERY

Harvard, 1923–1979

14

"I must confess that in Massachusetts I have found a 'stony-hearted stepmother.'"[1]

So wrote Cecilia, quoting the English essayist Thomas De Quincey, describing her first impressions on arriving in America in the fall of 1923. She had traveled from one Cambridge to another, and the contrast was stark.

She was a young woman of twenty-three, landing alone in a country she had never set foot in. After arriving by train in bustling Boston with her heavy bag and her violin case, she at first drew a bleak comparison to her homeland. In America, she lamented, one finds "a land where there is no spring, where summer comes in a sudden burst after the rigors of an icy winter . . . where there are no primroses, where the violets have no scent, where you will seek in vain for purple heather and golden gorse."[2]

No purple heather, no golden gorse. Just opportunity. It was the Roaring Twenties—anything and everything was possible. Wave after wave of immigrants had been drawn to the American Dream, to a land where it was said that neither caste nor class got in the way of success. The year Cecilia arrived, a publication in New York called *Time Magazine* put out its first issue; it was thirty-two pages and cost fifteen cents. In Los Angeles, the movie industry was flourishing. In the heartland between those cities, there were chain stores, mass-produced automobiles, electrical appliances. There was no country in the world like it.

No job in science for her in England, versus "the opportunity to live the life and follow the profession that I so much desired" in America—a clear example of the difference between the two societies.[3] Americans were curious, to the point of daring. Ex–fighter pilots bought old Curtiss Jenny biplanes and barnstormed across the continent, paving the way for Lindbergh's transatlantic flight in 1927. Radios were increasingly found in living rooms, and Model T Fords on the roads.

In America a century ago, science was celebrated, exploration embraced. Women, as L. J. had told Cecilia, could follow their ambitions and become real scientists. Cecilia had landed in the right place. "Yes," as she described it later, "we do things better here."[4]

Like all immigrants, Cecilia had her own set of hopes. Just as she had been a determined woman on a mission to learn in England's Cambridge, she was now on a mission to understand in this new Cambridge. With her stipend from Shapley she was technically a student, and so she was assigned a room in a Radcliffe dormitory. More contrasts. At Newnham, she had always had the proverbial room of one's own; at Harvard, she had roommates. Gone was her prized privacy. "I do not think that I had been undressed before anybody since I was a baby, and I suppose they found me ridiculously prudish," she remembered later. "When they found that I wore layer upon layer of underwear, they used to watch me disrobe with shrieks of incredulous delight."[5]

With her layers of clothes and her accent, she was unquestionably a stranger in a strange land. It could make for a little defensiveness. Her clothes, for example—the ones she bought before leaving for America with the small scholarship funds she worked so hard to earn—were different. On the back of a snapshot of herself standing in proud profile on the Harvard campus, Cecilia wrote to Betty Leaf: "This is me standing behind Everett House in my new clothes. They don't look as nice as they really *are*."

She missed Betty. She didn't know it yet, but the observatory was filled with offbeat personalities that would over time become like family. But in the meantime, it was not easy for naturally shy Cecilia. She continued on the back of the photo: "Be sure I am thinking of you, hence the pensive expression!"[6]

Dressed in her new clothes, her daily destination was a rambling brick complex a few blocks away from her dorm. On the back of a second photo for Betty, Cecilia wrote: "This is the great Harvard Observatory and the path that I walk up every morning. The dome at the end holds the 15-inch telescope. . . . Most of the Observatory buildings are to the right; you can see where I work through the bunch of fir trees."[7]

The observatory had been founded by the Corporation of Harvard College almost a century before Cecilia arrived. In October 1839, the Corporation bought two and a half acres of private land near the southeast corner of the campus. The house on the property had a cupola just large and strong enough to support a dome, but the observatory's telescopes were small. The original equipment was so inferior, in fact, that in March 1843 when "a comet of surpassing size and splendor appeared, and attracted intense interest . . . the Observatory could not satisfy the demand" for observations and information.[8]

Suddenly aware of how poorly equipped its observatory was, the Corporation raised additional funds and began a search for a new site. Because the site had to be close to the campus, yet still of sufficient size to house a large telescope, the search committee had to make the best of a less-than-ideal viewing location. The Corporation bought an old estate on Concord Avenue that "was for the most part country; and with no electric lights and no street cars or heavy trucks the situation was as favorable as could be desired."[9]

Determined to do it right this time, the Corporation authorized the purchase of a 15-inch equatorial telescope, known as the Great Refractor,

Cecilia at Harvard, 1923; photo for Betty Leaf

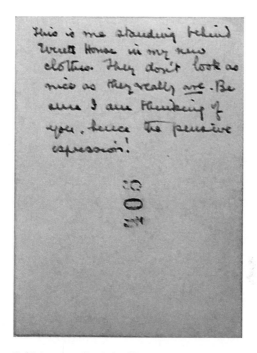

"This is me standing behind Everett House in my new
clothes. They don't look as nice as they really *are*."

the largest of its kind. Over two days in June 1847, the telescope was
placed on an eleven-ton granite pedestal and lowered into a specially prepared tower capped by a fourteen-ton hemispheric dome thirty feet in
diameter.

When completed, the new Harvard College Observatory was one of
the few observatories in America that could compete in stargazing power
with those in Europe. The labyrinthian structure, complete with a director's residence, looked very different from the rest of America's oldest
university. Situated on a hilltop, framed against a moonlit Cambridge
winter sky, the lights in the windows glowing deep into the night, this
was no library, no traditional classroom, no dormitory. Katherine Haramundanis, Cecilia's daughter, wrote about it years later through a child's

Cecilia's winding walk to work at the Great Refractor, 1923

eyes. She remembered the "spookily delightful spiral staircase" and "creaky dumbwaiter" (only a narrow spiral staircase connected the four stories of plate stacks, making a dumbwaiter, much like those found in old hotels, helpful to the staff). "There was also the dome of the 15-inch telescope, frigid in winter, with a chair on a tall scaffold that went squeakily up and down as the seated rider turned a large wheel in his lap. Beneath the circular rotunda of the telescope lay its great stone footings surrounded by a dusty catacomb."[10]

There was more than just dust in those catacombs. Cecilia would find hundreds of thousands of photographic plates stored in the observatory's underground chamber. Each plate contained spectra of numerous stars, generated by an objective prism. Virtually every star visible to the observatory's telescope was represented.

The plates had been developed, viewed, and classified years earlier by a smart, hard-working handful of women. Brilliant but quirky, the Harvard Observatory "computers," as they were known, would become folklore. Cecilia would not have known to describe it this way, but her arrival just as this group was completing its work was a turning point in the history of astronomy. The computers, with their classification system, had captured in the catacombs the mystery of the universe; Cecilia, with her knowledge of astrophysics, held the key to understanding it.

The computers had been hired by someone Cecilia would never meet, a man obsessed with collecting starlight. Edward Charles Pickering was the director for most of the observatory's first eighty years. A true Bostonian, he was born in Beacon Hill in 1846 and died in Cambridge in 1919. He was a thirty-one-year-old professor at MIT when he was chosen by Harvard president Charles Eliot to be the observatory's director in 1877. Old school astronomers were not thrilled that a physicist instead of a trained stargazer was named to be the head of the prestigious observatory. But President Eliot's primary job was to be the caretaker of Harvard's reputation, and he knew what he was doing.

From the very day he took over, Pickering dedicated himself to making sure that the Harvard Observatory would be the leading stellar observation post. He believed that science could best be advanced by obtaining enormous amounts of data to sift through. His goal was to create a "history of the sky" by attaching a spectroscope to a telescope in order to build a vast collection of stellar photographs, called spectrograms. His timing was perfect. So-called photographic dry plates—thin sheets of glass coated with emulsion to make them light sensitive—were just beginning to be mass produced.

Pickering's process worked well. Capturing large amounts of data, a task that had previously taken months, could now be done in a few hours. In addition, the spectroscope could "see" far more stars than the most highly trained astronomer with the very best eyes. "Something of romance was perhaps lost by the introduction of photographic methods," remarked the chronicler of the Harvard Observatory, "but the gain in efficiency was tremendous."[11]

The Harvard College Observatory, however, wasn't the only US observation post. There was growing competition. During Pickering's reign, larger telescopes were either under construction or on the horizon: the 36-inch refractor (1888) at the Lick Observatory on Mt. Hamilton just east of San Jose, the 40-inch (1895) at the Yerkes in Wisconsin, and especially the giant 60-inch (1908) and 100-inch (1917) telescopes at Mount Wilson northeast of Pasadena.

At first, Pickering was a one-man band. He invested his own capital—reportedly $100,000—and constructed his own equipment. He worked alone for several hours every clear night, ultimately making more than a million observations in total. To stay competitive, however, he knew he would need help. He began to take time out from stargazing to campaign for more funding, and eventually he hired assistants—all young men. But he was so dissatisfied with their productivity that he is said to have stormed out of the building one day "vowing that his Scottish maid could do a better job."[12] So, he hired her. Her name was Williamina Fleming. She was only twenty-one years old in 1878 when she followed her banker husband, James Orr Fleming, sixteen years her senior, from Dundee, Scotland, to Boston. James abandoned Williamina, pregnant with their son, a year after they arrived in the United States. She needed a job; Pickering needed a maid.

If Pickering had worked in England, could he have offered the same opportunity to his maid? It's hard to imagine that he could have. But Pickering must have known that he had more than just a domestic in his employ. Williamina had been a teacher in Scotland, and her craftsman father was the first to introduce daguerreotype photographs to the good people of Dundee. She was vivacious, and she clearly had an intellect to match his. Together they took the early stellar classification system of Angelo Secchi, director of the Roman College Observatory, and refined it into a sophisticated catalog system.

They were a most effective team because Williamina was ambitious, but not overly. Pickering wanted his assistants to gather data and not overthink it. Williamina did just that. When the president of Radcliffe, Elizabeth Agassiz, happened to visit the observatory in the spring of 1899, she overheard Williamina describing her work routine to a class of stu-

Williamina Fleming

dents. "It is impossible to reproduce the charm of the narrative as told by Mrs. Fleming," Agassiz recounted in her Radcliffe commencement address that same year. "The fitting of the blank photographic plate into the glass at evening, the setting of the telescope to the prescribed area over which it is to travel before daylight returns, the winding of the clock which is to control its motion, the examination of the plate in the morning, and the finding possibly [of] a new star."[13]

Williamina cataloged hundreds of novae, variable stars, and nebulae. Pickering published the discoveries in the *Annals of the Harvard College Observatory*. To his credit, Pickering told an 1898 conference at Harvard that Williamina had actually been the discoverer of seventy-nine new stars, "whereupon Mrs. Fleming was compelled by a spontaneous burst of applause to come forward and supplement the paper by responding to the questions elicited by it."[14]

In a personal journal she kept, Williamina described what life was like for a working girl in astronomy. Because she pored over the plates Monday through Saturday at the observatory, she had to set aside Sunday mornings for chores and laundry.[15] But she was nonetheless grateful to her boss; she managed to bring up her son completely on her own. She named him Edward Pickering Fleming.

Having seen how productive Williamina was, Pickering figured he could fulfill his goal of charting virtually the entire heavens by hiring smart women on the cheap. He set the official salary at twenty-five cents an hour, and the work schedule at seven hours a day, six days a week, with one month paid vacation. He envisioned the observatory as a data factory.

So, in 1888, when he wanted to improve and refine the Pickering/Fleming classification system, he went searching for another smart woman. He found Antonia Maury, the niece of Henry Draper, a major benefactor of the observatory. And she *was* smart—she read Virgil in the original Latin at age nine and graduated from Vassar with honors in astronomy, physics, and philosophy.

Alas, Antonia was too smart. She didn't just refine Pickering's homemade catalog—she set up her own independent system. Pickering watched in bewilderment, then in dismay, and finally in exasperation as Antonia "did the work as she herself thought best, instead of dutifully, and without questioning, following the direction of her superior."[16]

Tension filled the Harvard College Observatory. Pickering thought that "the detail Maury recorded for each star was a waste of time."[17] Antonia chafed—she wanted to use her classification system to explore stellar evolution. Pickering groused—he wanted results, and he wanted them now. It couldn't last. After four years of increasing discomfort, Antonia departed for a teaching job. Resisting the pull of trying to understand the heavens, however, was hard for her to do. She returned for a few months in 1908, and would routinely visit the observatory for the next decade. In 1918, she was appointed an adjunct professor at Harvard.[18]

Antonia Maury

Cecilia bonded with Antonia. They had a number of things in common. Both disdained fashion—Cecilia heard it said that Antonia would come to work "in one black stocking and one brown." And both had an un-apologetically deliberate way of working—Antonia "was always slowing things up by asking what it meant." Antonia even confided to Cecilia that she looked upon her as "the daughter she had always dreamed of having."

But it was the work, the pursuit of understanding, that was the strongest bond. "Miss Maury was a dreamer and a poet," Cecilia recalled later. "Many were the long talks that we had about the problems of stellar spectra. We both liked to work at night, and our discussions were painfully punctuated by insect bites, for she insisted on keeping the windows open and could not bear to kill the mosquitoes."[19]

Even Cecilia had her limits, however. Insects she could put up with if work was being done. Idle chatter was another matter. There were times when Antonia "just talked and talked and talked and talked. You couldn't do any work because she wanted to talk so much."[20]

With Antonia, Pickering had learned his lesson. His next hire was a dream come true. Annie Jump Cannon was the Goldilocks of observatory computers—smart enough to attack stellar classification with thoroughness and competence, yet not so driven as to gum up the works.

Annie was a perfect match for Pickering's needs. Most important, she got along with him. Her unquestioning style, "together with her reportedly charming personality, explains why she worked more smoothly with Pickering than did Maury," notes the historian Pamela Mack.[21] And physically, she was made for the task. "She had wonderful eyes," recalled Cecilia in an interview years later. "She could see things that very few people would recognize until she pointed it out."[22]

Annie's father was a well-to-do shipbuilder and Delaware state senator. Her mother was cultured and curious; she was fascinated by stars, early on pointing out various constellations to her daughter. As a child, Annie would go up into the attic of the family home, "open the trap-door and look over the tree-tops at the stars. She taught herself the constellations from crude charts in an old astronomical book, using a tallow candle for seeing the chart."[23]

Annie graduated from Wellesley College in 1884 with a degree in physics and then returned home. She was partially deaf, probably as a result of a bout with scarlet fever as a child. She was intellectually gifted; but stuck at home and unable to hear well, she was isolated from other people. She wrote in her personal journal: "I am sometimes very dissatisfied with life here. I do want to accomplish something so badly."[24]

Her mother's death in 1894 seems to have been the key that freed her. She got a job as an assistant in the physics department at Wellesley. A year later she won a "special student" place at Radcliffe that allowed her to study astronomy. Pickering saw those eyes and appreciated that mind; he hired her to take up where Antonia had left off, presuming—hoping—

Annie Jump Cannon

that she would be just an observer and not "be hampered by any preconceived theoretical ideas."[25]

Thrilled to finally have the chance to "accomplish something," Annie tore into the task. There is a 1930 black-and-white photograph of her in the Harvard Observatory files that perfectly captures this prodigiously productive woman: both hands gripping a photographic plate, her intensely focused eyes squinting inches away, wearing a formal suit, a string of pearls around her neck. She "reigned supreme," Cecilia wrote. "She would take an objective prism plate and would swiftly number the images with a pen. Then, with a recorder at her side, she would classify the spectra, speaking as fast as the recorder could write."[26]

Pity the poor recorder. With magnifying glass in hand, poring over 8 × 10-inch glass plates only a few millimeters thick, Annie could classify spectra into different types at an astonishing rate: three hundred stars per hour, or one every twelve seconds. Over the course of her career, she would inspect more than fifteen thousand photographic plates. And she did it with her own style. "She wore her hearing aid with an air," recalled

Henrietta Leavitt

Cecilia, "and made a virtue of necessity by unshipping it [from her ear] when she wanted to be undisturbed or to do concentrated work."[27]

The last of Pickering's "harem," as the computers were sometimes known, was Henrietta Leavitt. After graduating from Radcliffe in 1892, she worked at the observatory for a year, but only as a volunteer. Years later, she wrote to Pickering and asked about returning. She was dignified, formal, and classically educated, and, like Annie, she possessed the singular focus of the partially deaf. She was another perfect-for-Pickering observer, and the director knew it. To make sure that he got her, he bumped up her starting salary—from twenty-five to thirty cents an hour.

Like Antonia Maury, Henrietta was hooked on understanding what she was observing and classifying. Unlike Antonia though, she didn't complain; she simply discovered, quietly, on her own. While studying variable stars—stars that regularly change in brightness—in the collection known as the Magellanic Clouds, she noticed something intriguing. She discovered that she could relate the brightness of each star to the amount of time it took to cycle from dim to bright again. This finding was elegant in its simplicity, but no one had ever noticed it before.

The Harvard College Observatory computers. Cecilia is in the back row, second from left.

Astronomers would eventually use her discovery of the "period-luminosity relation" to measure a star's distance from earth.

Pickering's wise choice in hiring Henrietta had paid off. He knew that her discovery was an important one. He published the results in the *Harvard College Observatory Circular* of March 12, 1912—under his name. The report did begin, however, by noting that "the following statement . . . has been prepared by Miss Leavitt."[28]

The Harvard College Observatory family. Cecilia is the fifth adult from the right.

"Pickering chose his staff to work, not to think," observed Cecilia. And did they work! The team of computers produced nine volumes of stellar classifications, each volume containing more than 250 pages. In Cecilia's words, "The resulting catalog was a model of conciseness, consistency and accuracy."[29]

The computers didn't quite do it alone, however. There were supporting actors, a group of elderly assistants with their own idiosyncrasies. There was Louisa Wells, whom Cecilia remembered as "sitting at her desk marking stars on a plate, and then falling asleep and rubbing off all the marks with her nose"; and Edward King, who always admonished Cecilia that "one should never record the time of ending an exposure until the shutter had actually been closed; one might die in the interval, and the record would then be inaccurate."[30]

And then there was Frank Bowie, the night assistant. He was also deeply absorbed in Cecilia's calculations, but in a different way. By day,

Cecilia at Harvard, 1924

he managed a numbers game in Boston. He believed that "the Right Ascension and Declination of a newly-discovered comet was considered a 'lucky number' and was duly played. On one occasion it paid off handsomely, and enhanced the faith of the underworld in the power of the stars."[31]

It was, to be sure, an oddball cast of characters, and a setting out of Dickens—a forbidding domed brick building on a hilltop set apart from the swirl of students in Harvard Square. But for Cecilia, far from her mother, her siblings, her best friend Betty, and her beloved England, the observatory and its staff gradually became home and family. In time, she "found at the observatory the companionship and social status others achieved through kinship."[32]

Cecilia came to know only two of the computers: Antonia Maury and Annie Jump Cannon. Williamina Fleming died in 1911. Henrietta Leavitt's days were cut short by cancer; she died in 1921 at the age of fifty-three. But Cecilia certainly knew of her: "I think she was the most brilliant of all the women [at Harvard]," she said.[33]

There was also a historical link, a sweet connection between past and present. "I never saw Pickering, never knew Miss Leavitt," Cecilia wrote years later, "though their shadows could still be discerned. I heard tell that Miss Leavitt's lamp was still to be seen burning in the night, that her spirit still haunted the plate stacks. I suspect that some credulous soul (and there were such in those days) had seen me from afar, burning the midnight oil. Shapley had given me the desk at which she used to work."[34]

15

"I had left the world of dreams and stepped into reality." So wrote Cecilia when she described her transition from the Old World to the New. "Abstract study was a thing of the past; now I was moving among the stars."[1]

It was time to put all that Cavendish Lab learning to work. Her first task was to get to know the Harvard College Observatory's new director, Harlow Shapley. They had met when L. J. Comrie had introduced Cecilia to him in London, and they had corresponded about her coming to Harvard. But she didn't know much about his approach to science. Would he be as driven to understand the heavens as she was? She would not be disappointed. Years later, she noted that "his mind traveled about the stellar universe as in familiar country."[2]

Shapley was a country boy. He was born on a farm in Nashville, Missouri, and before he was even in his teens he was on his way down a ne'er-do-well path, dropping out of school after fifth grade. He studied at home and scrounged a job as a cub newspaper reporter covering local crime stories. He had a restless mind, however (he would maintain a lifelong interest in myrmecology, the study of ants). He ended up returning to high school and finishing the six-year curriculum in just two years. He graduated as class valedictorian.[3]

At twenty-two, he was older than his classmates when he enrolled at the University of Missouri as a journalism major. Still restless, when he learned that the School of Journalism wouldn't open for another year, he grabbed a course catalog and vowed to study the first course on the list. That was archaeology. He couldn't pronounce it, so he went on to the next one, astronomy. He took to it with a zeal as burning hot as the stars he studied.

Upon graduation he won a fellowship to Princeton, where he studied under Henry Norris Russell, the Princeton Observatory's rising star. The astronomical community was small; everyone knew what everyone else was working on. Shapley's star too began to rise when he calibrated Henrietta Leavitt's period-luminosity relationship to show that the Milky Way was considerably larger than previously thought, and that our sun is in one of its arms, not the center. This discovery was seen as extending Copernicus's insight: not only is the earth not the center of the solar system, but it is not the center of our galaxy, much less of the universe.

It was while working at California's Mount Wilson Observatory, located on top of a 5,700-foot peak in the San Gabriel Mountains near Pasadena, that Shapley got his big career break. Edward Pickering had ruled the Harvard College Observatory for forty-three years, working until the day he died of pneumonia on January 6, 1919, at the age of seventy-two. For the next two years, the observatory was without a director while the president of Harvard, Abbot Lawrence Lowell, debated with various astronomers and with himself over whom to appoint. The process was a stellar example of academic palace intrigue.

There were three recommended candidates: Shapley; Frank Schlesinger, director of the Allegheny Observatory at the University of Pittsburgh; and Henry Norris Russell, director of the Princeton Observatory, who had been Shapley's thesis advisor. Shapley's boss at Mount Wilson, George Ellery Hale, had forwarded to President Lowell a letter of recommendation for Shapley from a fellow astronomer, along with his own verbal recommendation. In handwritten notes at the bottom of the letter, Lowell wrote that "Hale tells me that on the whole he puts Shapley first. . . . He is about 30, very brilliant, but he is not sure how well he could manage men."[4]

Apparently Shapley had the scientific credentials, but his management skills were suspect. To assess this more personal side, Lowell privately arranged for Matthew Luce, Harvard's regent, to send tickets to Shapley and his wife for the 1920 Rose Bowl, which Luce would be attending with the Harvard team. By the end of the game, the Crimson, the odds-on underdog, had eked out a 7–6 win over Oregon. Shapley also eked out a win. Luce, who had been able to spend some time with him, reported back that Shapley had an "agreeable and serious personality. Attitude of his fellow workers toward him good."[5]

"As luck would have it," Shapley later recalled, "the Harvard team won. Nobody expected them to win because they were gentlemen, not football brutes, but they did. . . . So back east they decided to take a gamble on me." Actually, not quite yet. The underdog gentleman still faced some unexpected headwinds. Hale's original unqualified praise grew a bit fainter in a subsequent letter to Lowell. Hale wrote that he would be unwilling to turn over the directorship of Mount Wilson to Shapley because "he has not yet reached complete maturity"; nevertheless, "I really believe he would prove a great success at Harvard."[6]

Not good enough for Mount Wilson but fine for Harvard—not exactly what President Lowell wanted to hear. He ordered more sizing up. In the spring of 1920, Shapley faced off with the astronomer Heber D. Curtis in the National Academy of Sciences' "great debate" on the scale of the universe. On hearing that Shapley did not do well, and with Schlesinger having accepted the directorship of the Yale University Observatory, Lowell quickly notified Henry Norris Russell at Princeton that he was under active consideration for the Harvard position. Russell sent back word proposing that perhaps a team effort was the way to go. "Shapley couldn't swing the thing alone, I am convinced of that," he wrote to Hale. "But he would make a bully second."[7]

Lowell was no doubt concerned about too many cooks. He telegrammed Russell with the offer of sole directorship. Russell turned him down but then recommended Shapley! Lowell was caught now in a swirl of conflicting signals. To the rescue came Hale. Hale wrote Lowell that he would be willing to give Shapley a year's leave-of-absence from Mount Wilson so that Lowell could try him out to see if his management skills were up to the position.

And so it was that Harlow Shapley started work at the Harvard College Observatory in April of 1921 as an "observer." It was officially a year's probation, but Shapley was appointed permanent director just six months later.

Two years into the job, Shapley must have beamed when he saw Cecilia walk through the door. He may not have envisioned forming a "Shapley's harem," but he certainly viewed Cecilia as a worthy member of his team of new "computers" to carry on the Pickering legacy. He initially gave her the courtesy of asking what she wanted to work on. But before she could answer, he pressed on, saying that he thought she would be perfect for continuing Henrietta Leavitt's work on standard photometry. Cecilia had other ideas. "I was in a different position from the other girls," as she described it. "They were employed to do a job, but I was on a Fellowship, so I was independent and had no obligations."[8]

She was correct. Technically, she was not an employee, so Shapley couldn't tell her what to do. She was a smart, hard-working young woman, living on a subsistence income with an office in the observatory—the very definition of an observatory computer—but she was not under his thumb. That realization gave Shapley pause. He then asked, delicately, what exactly she *did* want to do. Cecilia responded as she always did— directly, to the point. She said she wished to do what E. A. Milne had suggested: test Meghnad Saha's theory of stellar composition.

The observatory's million photographs held a vast storehouse of scientific data. But without interpretation, it was as if a treasure trove were

Harlow Shapley

held under lock and key. What Cecilia was proposing was a way to unlock it: apply her Cavendish Lab knowledge of physical chemistry to the Harvard College Observatory's collection of stellar spectra. She wanted to bring astrophysics to Harvard.

Shapley may have initially misjudged Cecilia, but he did recognize the value of combining the Cav Lab with the HCO. As Cecilia described it in an interview years later, Shapley told her, "'All right, go ahead. There are the plate stacks.' So I was left just to sink or swim. There wasn't anybody to help because it was a subject nobody knew about."[9]

Shapley, however, had a much more ambitious agenda for himself and for the observatory than just caretaking a data factory. His mentor, Russell, was a professor of *astronomy* at Princeton. Even his alma mater, the University of Missouri, offered a course in *astronomy*. Shapley thought the Harvard Observatory should be an integral part of Harvard University,

Harvard College Observatory, 1925

not just a research outpost. He wanted to be the founder of nothing less than the Harvard Department of Astronomy.

And if he couldn't get Cecilia to join the ranks of Harvard Observatory computers, perhaps he could use her to further his astronomy ambition. He prevailed on her to write up her findings in the form of a thesis and thus become the first doctoral student in astronomy.

At first, Cecilia was nonplussed. She had graduated from Cambridge; she didn't see the need. "I was not much interested; I thought that no degree could be higher than the one I had received from Cambridge University (even [if] in those days, before the admission of women, it was only the 'Title of a Degree')."[10]

Shapley was persuasive. He told her she didn't have to take any actual astronomy courses (there weren't any!); all she had to do was write a PhD thesis. Cecilia knew Shapley wanted to use her as "the thin edge of the wedge" to jump-start an astronomy department.[11] She agreed. Neither of them realized at the time what a momentous decision had been made.

The observatory had two libraries—one containing printed works, and one all of those photographic plates. Cecilia attacked the written word first. "Through the library I went, shelf by shelf, arranging its contents in the pigeonholes of my mind." She pored over every book and every periodical, systematically indexing their contents on note cards. It was here, in a traditional but focused collection of published research, "where the dry bones of astronomical knowledge were stored."[12]

It is interesting that she used the metaphor of pigeonholes. Betty Leaf had used the same phrase during their first year at Cambridge to describe how Cecilia's mind worked. Cecilia kept a snapshot of Betty in her wallet, from seemingly years ago and an ocean away.[13] She surely felt a pang of homesickness until she put the photo away and turned her eye to other photographs.

Cecilia banished thoughts of home by going after the collection of stellar plates with the same intensity she showed for the regular library. Ironically, her early botany training proved valuable. William Bateson and Agnes Arber had shown her the importance of the systematic classification of plants. Working as if she were a stellar botanist, Cecilia applied the same principle to the plates, viewing her task as "ranging over the astronomical photographs, collecting and classifying the celestial flora."[14]

There were hundreds of thousands of stellar photographs; analyzing them would be hopeless without a systematic approach. Cecilia created a set of log books, with her name carefully recorded on the flyleaf, each one a comprehensive recording of the star she was studying, crafted in the same careful handwriting she had used to correspond with Shapley. "A look at her log books from the photographic plate stacks shows a person who hit the ground running as she searched the cumbersome and voluminous archive," her daughter, Katherine, remarked.[15]

The work of cataloging that "voluminous archive" had already been largely done. Pickering's computers had swept the sky clean by the time Cecilia took over Henrietta's desk. Over more than twenty-five years,

Annie Jump Cannon alone had classified 350,000 spectra. But that's all Annie did. "She had amazing visual recall, but it was not based on reasoning," Cecilia noted. "She did not think about the spectra as she classified them—she simply recognized them." It was as if Annie had built a massive library and stocked it with an enormous number of books but then never read a single one of them. Cecilia could not help but "wonder how anyone who had worked with stellar spectra for so long could have refrained from drawing any conclusions from them."[16]

There were two reasons. First of all, it was not Annie's job to draw conclusions. She had been hired to use her eyes. But more important, Annie did not have the tools for exploration that Cecilia had. She had not been trained in the rigorous environment of the Cavendish Lab. She had not learned from Niels Bohr about how electrons orbit a nucleus. She had not had a determination to understand honed and hardened by a Nobel laureate singling her out as the only woman in the physics class.

Only later did Cecilia consider that Annie might well have resented a young student "presumptuous enough to attempt to interpret the spectra that had been her preserve for many years." Shapley had also recognized that Cecilia was taking a risk; he asked her once if she realized "how easily Miss Cannon could throw a monkey-wrench into the works for you?"[17]

Cecilia described Annie as "extraordinarily kind" to her, but it was another newcomer, Adelaide Ames, with whom she forged a quick and strong friendship. An only child and Army brat raised in Boston, Adelaide graduated from Vassar in 1922 and then followed a path to the observatory similar to that of the director. Just as Shapley was a cub reporter turned astronomer, Adelaide too was torn between journalism and science. Shapley made up her mind for her; he hired her as his assistant. Adelaide was quoted in an article as saying, "A job in astronomy was offered me and none in newspaper work." The article continued: "Whatever her future might have been in journalism, she proceeded to make her mark in astronomy. In collaboration with Dr. Shapley, Miss Ames published

Adelaide Ames

several volumes of astronomical observations in relation to new galaxies which are among the most complete published."[18]

The newspaper may have pumped up Adelaide's role a bit; "collaboration" surely was not the way the director would have described it. And Adelaide knew her place. "I collect only the facts," she once said. "The theories are Dr. Shapley's."[19]

Both in their early twenties; both working in an intense hothouse of research with an older, quirky group of people on the periphery of the campus; both struggling to make ends meet—little wonder that Cecilia and Adelaide immediately gravitated toward each other. "She was young, lovely, intensely vital," Cecilia recalled. "In my first year at Harvard we had been inseparable; they used to call us 'the Heavenly Twins.'"[20]

Shapley got away with offering small salaries because he managed to create in himself another form of currency. Sometimes the alternative currency would be deducted; if a computer or assistant were not at her desk when he came out of his office and walked around, he would leave a note. Sometimes it would be added; he would regularly pause in his daily stroll to remind the staff how important their particular job was. Acknowledging his reliance on women workers, Shapley, according to Cecilia, "measured his projects in 'girl-hours.'"[21]

"Everyone adored him—the older women and young girls who were soon added to the team," Cecilia remembered. "Adelaide and I called him 'the Dear Director,' and soon he was affectionately known as 'the D.D.'" The two of them used to say jokingly that "he had found a Dear Little Observatory, and intended to leave it a Great Institution."[22]

The "D.D.," however, might just as easily have stood for Dear Dictator. It was Shapley's shop, and as for Cecilia, "he never forgot, or let me forget, that he was the Director of the Observatory." With her newfound independence, she was feeling more and more like a woman, and she found Harlow to be boyishly charming, "running upstairs two steps at a time, pushing his soft sandy hair off his forehead." He could also be "vain and vindictive," however, and he "kept his distance"; even after knowing Cecilia for more than fifty years, he never called her by her first name. Still, "in those days I worshipped Dr. Shapley," Cecilia recalled later, describing herself as a twenty-five-year-old. "I would have gladly died for him, I think."[23]

The small compensation—the stipend plus the boss's motive-driven words of encouragement—hardly mattered, because the opportunity was priceless. This was it—the chance to be more than a schoolteacher. "I had the run of the Harvard plates, I could use the Harvard telescopes (a dubious boon, this, in the climate of Cambridge), and I had the library at my fingertips."[24]

From Newton to Einstein to Eddington to Saha, there were so many theories; but no one had yet discovered what the stars were made of. To Cecilia the search was thrilling. "The history of science is a history of delight in first-seens, first-postulateds, first-came-upons," writes Kay Redfield Jamison in her book *Exuberance*. "It is a history of high pleasure in the hunt and exultation in the netting."[25]

This hunt was on: Cambridge training meets Harvard data. Cecilia never described her feelings at that moment, but she had to know that she had a shot at a major discovery. "I saw in the stars a chance to observe phenomena beyond terrestrial scope. Nothing seemed impossible in those early days; we were going to understand everything tomorrow."[26]

Cecilia's extraordinary level of energy was now finally unleashed. At Cambridge, if she had wanted to study after 11 P.M., it had to be done in bed by candlelight. At Harvard, she could come and go as she pleased, work all night if she wanted to. "When she set herself a task, she was indefatigable," according to her daughter, Katherine. "Her powers of concentration were so great that she could work for hours without stopping."[27]

At first it was too much. Cecilia was trained in modern atomic physics, but when she began examining Annie Cannon's quarter million observations, she found the spectrum of starlight on any given plate to be little more than "tiny parallel smears."[28] How was she to apply atomic theory to a smear?

Despair comes in different forms to different people. To a scientist, it comes as bewilderment. Months and months of time; packs and packs of cigarettes. No progress. Cecilia despaired. The tiny smears simply would not reveal their secrets. Shapley watched and waited. For a man whose career depended on results, it must have been excruciating. He had to have had moments of doubt about the wisdom of offering precious plate access to this admittedly hard-working young woman with an increasingly edgy personality.

Late one night it bubbled over. She could hear his footsteps as he jogged across the courtyard from his residence to the Brick Building. He approached Cecilia, sitting as usual at Henrietta's desk, plates spread out under the lamplight. "Don't you think you should publish something?" he asked. "To give some evidence of the work you're doing."

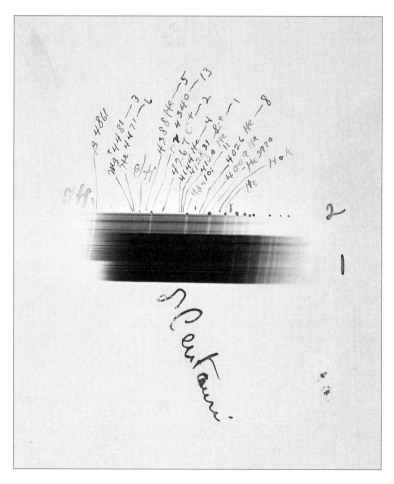

Spectroscopic smears

"No!" she snapped. "I should regard it as a confession of failure." Shapley backed away.[29]

She redoubled her already redoubled efforts. It was intense, and made even more difficult because there was no one around to lean on. Her observatory colleagues were a close-knit group, especially Adelaide, but her closest friends and family members were far away. L. J. Comrie was at Swarthmore; Betty Leaf was back in London. Her younger brother and sister were also far away: Leonora was still in school, pursuing an architecture degree; Humfry was off on archaeological digs in Greece. From time to time she would write to her mother, but the letters were no substitute for conversation. Emma was perceptive enough to read between the lines. Unbeknownst to Cecilia, Emma wrote Shapley a heartfelt letter expressing concern about her daughter: "She is healthy, but not really a strong person, and lives largely on her enthusiasms, and while I delight to think of her doing the work she loves, I cannot help being anxious."[30]

Emma was right to worry. The same force that enabled Cecilia to endure Rutherford's derisive remarks, Searle's caustic corset comments, and Bateson's dismissal of her enthusiasm for research was at work again. It was an impatience with the ordinary—with sleep, meals, even friendships and family—that had driven her as far back as she could remember.

"At a very early age," Cecilia wrote later, "I made up my mind to do research, and was seized with panic at the thought that everything might be found out before I was old enough to begin!"[31] What she was understanding was that her learning was not complete. She still had to realize that she was not "going to understand everything tomorrow." She was smart, but she wasn't knowledgeable enough yet. Many great leaps of discovery are made with painstakingly small steps. True progress is made slowly, with many unexpected turns. She took comfort in the thought that "knowledge comes to us gradually. If we were faced with the full complexity of the facts at the beginning of our search, we should be so bewildered that we might well give up in despair."[32]

That is when the smears of stellar spectra started, ever so slowly, to come into focus. As she became more familiar with Annie's classification system, Cecilia laboriously developed a method for identifying and estimating the intensities of different lines in the spectra (each pattern of lines representing a different element). The first element whose lines she made sense of was silicon ("still one of my favorite atoms"). Bohr had

Cecilia at her desk in the observatory

theorized that if an atom absorbed enough energy, one or more electrons would jump from a lower to a higher energy level and then, with
more energy, ionization would occur: electrons would leave the atom
altogether. A spectrogram of the ionized atom might appear to be that of
a different element (the spectrum of an ionized atom that has lost one
electron somewhat resembles that of the element to the left in the periodic table). But it isn't; it's the same element, just in an ionized state.

Keeping that idea in mind, Cecilia thought she could make out four
stages of silicon ionization in the photographic plates. "Finally some light
dawned in the darkness," she wrote later. She took a small calculated leap,
relating ionization levels to temperatures. "I made my first determination of the temperatures of the hotter stars."

Shapley was beside himself. Her training and his data were at last
paying off. He told her to write up her results—right away! It was her

first paper on stellar spectra. She wanted to sign it as C. H. Payne. Shapley stared at her. "Are you ashamed of being a woman?" he asked.[33]

The question jolted her. Cecilia had always thought of herself as being "on equal terms with any astronomer in the world." She was a scientist and a scholar, and "neither of these words has a gender."[34] But she also wanted to be taken seriously. And the memory of having to sit in the front row simply because she was a woman was still fresh. A rueful smile and a subtle nod—she would let the words speak for themselves. She signed the paper Cecilia H. Payne.

She was in gear now. The spectrograms were increasingly recognizable, making more and more sense. She could see more lines than before, see traces of more elements. More research papers flowed. It occurred to her that she could use the characteristics of the spectral lines to understand the ionization potential of different elements. She brought the idea to Shapley; he immediately understood where she was headed. If true, it would be another step taken toward identifying what stars were made of. He told her to submit it to the prestigious *Proceedings of the National Academy of Sciences*. She then told him that the deadline was the next day.

"Write it up at once!" he exclaimed. "And I'll type it for you."

Cecilia had had so many conversations with Shapley since she landed at Harvard. She flattered herself that he enjoyed her company because of her obvious intelligence. And when Shapley had taken equal pleasure in conversations with other people in the observatory, she admitted she was jealous. This night she had him to herself. She hadn't the heart to tell him she was an excellent typist; she knew he just wanted to participate. "What a glorious evening," she recalled. "I wrote, he typed, far into the night."[35]

Though Shapley was excited, this particular paper was relatively small in scope and import. It was the type of paper that other scientists in the field would read, be mildly impressed by, and then file away in a folder of other interesting articles. The big idea—the breakthrough to momentous discovery—was still a few months away.

Ironically, when the breakthrough came, the reaction would not be excitement but doubt. Had she known the skepticism that would greet her later, it might have tempered the "kind of ecstasy" she felt at the moment. But for now, she was an eager grad student putting her freshly typed paper into the mail at three in the morning on a cool Cambridge night in May of 1924. She was high.

"I walked back to my room in the dormitory in a dream. My feet did not seem to touch the ground. 'I never knew before,' I thought, 'what it means to walk on air.'"[36]

16

"Menzel has come."

Shapley had called Cecilia into his office. She "found him looking rueful and apologetic."[1] There was good reason.

Princeton's Henry Norris Russell had long considered the Harvard Observatory a "land of settled habits." For sure, data—*good* data—had been collected over the past forty years. But all those stellar photographs just sat there, fossils beautifully preserved deep in the observatory's catacombs. Russell had urged Shapley to get on with it. "If I had to run the place," Russell had written Shapley before Cecilia arrived, "I would plan to draw in sharply on the large routine jobs." Enough collecting, he urged—turn the staff loose to conduct "investigations on specific problems—large problems."[2]

Shapley held the job that he knew his mentor, Russell, had spurned—just the ingredients for insecurity. He gave a nod to Russell's suggestions, and he asked Russell to become an external advisor to the Harvard staff. But in Shapley's mind, he, Harlow Shapley, would not be credited with molding the Harvard Observatory into a

Great Institution if he were seen as Russell's puppet. He intended to preserve Pickering's data factory.

Russell at the time was fast becoming the dean of American astronomy. Which meant that if there was even the whiff of discovery in
the air, he would smell it. Like Cecilia, he was intrigued by Meghnad
Saha's theoretical linking of temperature and pressure to the composition of stars. He also coveted all the Harvard stellar data. Unaware of
the work of Harlow's promising young woman doctoral candidate,
Russell sent his best student, Donald Menzel, to Cambridge so that he
could use the Harvard plates to investigate exactly the same questions
that Cecilia was working on.

Cecilia listened to Shapley's evasive explanation for why Menzel was
on the scene with puzzlement but not panic. Although she admitted that
at the time she jealously guarded the chance for discovery, she would later
come to conclude that "a problem does not belong to me, or to my team,
or to my Observatory, or to my country; it belongs to the world."[3] Menzel
was highly trained in laboratory spectroscopy and physics; she was a
graduate of the Cavendish Lab with its intense focus on the physics of
the nucleus. A combined effort would have been extraordinarily powerful.

It was not to be. To Harlow's way of thinking, it was the Shapley Observatory now. He was taking a chance on Cecilia; glory, if it came, was
not something to be shared. It all fed into his personal philosophy that
people do their best work when they're miserable. No one could earn a
doctoral degree, he believed, unless they suffered a nervous breakdown
in the process. "Work," he told Cecilia, "kills the pain."[4]

An agreement emerged. Menzel would concentrate on the spectral
lines of neutral metals, which are found primarily in stars of relatively
low temperature. Cecilia would focus on hotter stars. Russell congratu-
lated Shapley on finding and hiring Cecilia, but he was not entirely pleased
with the research overlap. He later told Shapley that if he had known of
Cecilia's parallel interest, "I should have set Menzel at something else."[5]
Cecilia pressed on, alone.

As if Cecilia Payne versus the secrets of the universe were not daunting enough, now the element of competition had been added to the mix. She would have preferred to collaborate; but if it were to be a race, so be it. "Competition is as ancient as the hunt," writes author Kay Redfield Jamison. "The same fire that rouses the thrill of pursuit is kept kindled by the joy of victory. There is pleasure in the run, of course, but the high glory is in being first across the finish line."[6]

Cecilia knew that speed and passion were essential to be first across the finish line. Discovery tantalizingly within reach, competition bearing down—it was intoxicating. "I almost worked night and day without stopping," Cecilia told a colleague later. "It was marvellous."[7]

Contemporaries of Cecilia, from Cambridge to Harvard, painted a similar picture when describing her: the woman could really focus. Distractions? There weren't any when she was locked into trying to understand something. After the meeting with Shapley, she worked even longer hours. She may have thought of the words of Henrietta Leavitt, a woman she had never met but whose presence she felt as she worked at Henrietta's desk. Referring to Beta Lyrae, a multiple star system in the constellation of Lyra, Leavitt once remarked that "we shall never understand it until we find a way to send up a net and *fetch the thing down!*"[8]

Cecilia had precisely the net Henrietta had imagined. With astrophysics in mind and the observatory's plates in hand, she felt certain that she could fetch the stars. It was self-confidence rooted in her particular branch of science. She was an astronomer, and "astronomers are incorrigible optimists," as she described it years later. "They peer up through a turbulent ocean of atmosphere at the stars and galaxies, forever inaccessible. They speak of million degree temperatures, of densities smaller than our lowest vacuum; they study light that left its source two hundred million years ago. From a fleeting glimpse, they reconstruct a whole history."[9]

Physics applied to the study of stars—it would prove to be a powerful tool. There were at the time only a few people in the global scientific

Meghnad Saha

community who were focusing on astrophysics in the same way as Cecilia. One of them was Meghnad Saha.

Saha was born in October 1893 in Seoratali, a village in the middle of what is now Bangladesh. He was the fifth child in a family of eight. Growing up as a member of a low caste in India, Saha washed dishes to pay for room and board at middle school. He excelled in math; he also loved ancient history and poetry and languages and archaeology.

In his spare time as a graduate lecturer in physics at the University College of Science in Calcutta, he read everything he could find about astronomy and physics. He read the *Popular History of Astronomy during the Nineteenth Century* by the astronomer Agnes Clerke. He read Bohr's papers on how an element's electrons could make quantum leaps to produce different spectral features.

He read, and he thought. And what he thought about would change the worlds of physics and astronomy. "It was while pondering over the problems of astrophysics, and teaching thermodynamics and spectroscopy . . . that the theory of thermal ionization took a definite shape in my mind."[10]

In 1919, the year Cecilia entered Cambridge, Saha wrote a paper, as part of his thesis, that involved an analysis of the Harvard classification of stellar spectra.[11] The observatory at Harvard had found that almost every observable star could be classified into one of seven discrete categories on the basis of its spectrum—categories that were suspected to indicate the star's temperature. It was called the Harvard Sequence— O, B, A, F, G, K, M—with O as the hottest (they are rare, 1 in 3,000,000) and M as the coolest (the most common stars). Our sun, for example, is a star in the G part of the sequence. An astronomical wag, a man no doubt, came up with a mnemonic to remember the sequence: Oh Be A Fine Girl, Kiss Me.

For a decade, astronomers had struggled to provide an answer for why all stars fit into such a neat classification, and for how a star's temperature affected its spectrum. Saha, with his growing knowledge of physical chemistry, atomic physics, and thermodynamics, was able to produce an equation that related a star's temperature and pressure to the spectrum it produced, thereby linking the Harvard sequence to the physical characteristics of all stars. In other words, "Saha was the first to link the structure of atoms to the appearance of their spectral fingerprints."[12]

Saha knew that there existed raw data that could be used to test the validity of his theory and calibrate the temperature sequence—data, for example, like that preserved in the glass spectroscopic plate collection at the Harvard Observatory—but he didn't have access to it. As a result, he had no idea that a young British woman astronomer, working in the mid-1920s at that very same observatory, would soon use his theory to decipher the dark lines on those very same glass plates.

To Cecilia, it was a team sport. Meghnad Saha was in Calcutta devising theories utilizing physical chemistry; E. A. Milne and Ralph H. Fowler were in England working on similar questions of atomic composition using more established techniques; and she was in America poring over stellar spectra preserved on thousands of photographic plates. If she could take the efforts of Milne and Fowler to extend Saha's insights and then apply them to the observatory's trove of data, she could show what stars are made of. What was seventy-two hours of nonstop work when discovery was in the air?

Not that the men of science didn't have a prevailing view of what composed the solar system and beyond. Eddington's equations describing how stars were structured worked well as long as the heavy elements—silicon, magnesium, aluminum, oxygen, iron—were the dominant substances. Astronomers, therefore, assumed that the principle of uniformity held throughout the universe. Temperatures might differ, but the composition of the sun and all the planets was deemed to be the same as that of earth. The American physicist Henry Rowland summed up the prevailing assumption in 1890. He speculated that "were the whole Earth heated to the temperature of the Sun, its spectrum would probably resemble that of the Sun very closely."[13]

It was a neat, relatively uncomplicated theory. The equations that Eddington formulated—the foundation of his seminal work, *The Internal Constitution of the Stars*—contained only two constants, which "could be manipulated to fit neatly with observed data." As a result, Eddington's work "provided a convincing and consistent picture of the physics of stellar interiors." Contradicting his assumption of uniformity was a daunting task, for "Eddington's powers were unmatched in that day, and so was his influence."[14]

And contradicting Eddington, as well as the other men of science of the time, was where Cecilia was headed. Her early training from William Bateson and Agnes Arber in classifying plants served her well. She had set up her own system of analyzing spectral lines, and then had spent hours studying the plates—her "celestial flora," as she had described it—quantitatively matching the spectral smears to what Saha's equations predicted. From her training, she knew what to look for. She knew that a given element's atomic structure would dictate where along a temperature range that element would absorb light—and thus be visible—and then where along the range it would suddenly stop absorbing light and become invisible again.

At first, the elements she studied behaved pretty much as predicted. Silicon, and then carbon, fit nicely into their predicted share of a star's composition, matching their abundances on earth and staying consistent from star to star. It was an exciting revelation—the spectral lines were showing that the atomic composition of all stars, no matter how hot or cool, was the same. But Cecilia was very much an astronomical archaeologist—the dig was as important as the uncovering. She saw the plates as "bones to be assembled and clothed with the flesh that would present the stars as complete individuals."[15]

The composition of the stars might be the same, but what exactly *was* that composition? Was it the same as that of the earth and the planets, or was it different? She kept digging, and that is when the trouble started. As she worked her way down the periodic table, helium—a relatively simple element with two electrons—was not behaving. The intensity of the lines across the entire range of spectral types was showing the element to be far more abundant than it should be; in fact, a thousand times more abundant.

But her data on helium was nothing compared with what she found when she focused on hydrogen, the simplest of all elements—one electron circling a single proton. Cecilia's calculations, based on Saha's equation, indicated that in the hottest stars almost all of hydrogen's atoms would be stripped of their electrons. Only a tiny fraction of the hydrogen atoms would retain their electrons and produce spectral lines. And yet this tiny fraction was producing incredibly strong lines. There was only one way that could happen: the tiny fraction had to be a tiny fraction of a huge number.[16] Hydrogen was showing itself to be a *million* times more abundant than predicted.

Cecilia had to know she was headed for confrontation. At the time, "the possibility that hydrogen was the primary constituent of the universe was not a welcome thought at all," notes the historian David DeVorkin. "Even though hydrogen was the most persistent line feature in the spectra of the stars, and sometimes the most prominent, astronomers felt strongly that it could not be a major constituent of the stars."[17]

Cecilia then did what most scientists do when their results are non-sensical: she kept the results to herself until she could figure out what was wrong. There had to be an error—hydrogen was so off the charts—and if she worked hard enough, she'd find it. Though in a self-described state of "total bewilderment," she gamely went about checking and re-checking her results.

Periodically, Shapley would set up tables in the observatory's Phillips Library in the form of a square and invite the staff for tea. He referred to these gab sessions as "Hollow Squares," but others within the observatory called them "Harlow Squares."[18] Cecilia loved the Squares, for it was in these sessions that she would talk shop with visiting astronomers—a distraction from the confounding plates. Gender stereotypes had no place here. The visitors, including Knut Lundmark, head of the Lund Observatory in Sweden; E. A. Milne, her friend and mentor at Cambridge; Albrecht Unsöld, a German astrophysicist and expert in spectroscopic analysis; and Otto Struve, a Russian-American astronomer who became head of the Yerkes Observatory, treated her as their equals. After each Hollow Square, they would continue talking. "How we argued," she recalled later. "How we walked about the streets and sat talking in restaurants until the manager turned off the lights in despair!"[19]

Equals, yes, but there was nonetheless a first among them. "Russell has come!" was the excited word that bounced around the observatory when Henry Norris Russell, director of the Princeton University Observatory, came to Harvard. The process of discovery would wait. Cecilia raced for her spot in the Square. "We young people put aside all work and sat at his feet. Henry Norris Russell was a formidable figure, tall and lean, endlessly voluble, speaking with the voice of authority."[20]

Russell was at home in front of an audience. Born on Long Island in 1877, his role model was his father, a Presbyterian minister whose parish-

Henry Norris Russell

ioners included Theodore Roosevelt. His affinity for nature's uniformity may have sprung from the consistency of his personal life. He enrolled in a Princeton prep school at twelve and spent his entire career at the university, where he lived in the same house from 1890 to 1957. He graduated from Princeton in 1897 at the head of his class—high-strung, fast-talking, unable to relax, barely able to rest. He was like a clock that stops only when it winds down.

"We drank until he ran dry," recalled Cecilia. "After several hours would come a time when his words flowed more and more slowly. Finally one would hear him murmur: 'Mustn't go to sleep,' and then lapse into brief catnaps, punctuated with more words of wisdom." People joked that "he was the only scientist who had been known to go to sleep during one of his own lectures."[21]

Still, the spectrograms kept calling. Late at night she could be seen staring at the plates and her calculations, trying to find the error. It was grinding, frustrating work—hydrogen continued to claim an absurdly large piece of the pie—and it was gradually taking a toll on her. It was possible that she was on to something, a major discovery, and it was scary—a twenty-five-year-old woman trying to pry out secrets that nature had been keeping hidden. It was also all-consuming; a day's stopping point was impossible to find.

Shapley noticed. The "Dear Director" knew she needed a break. He told her—ordered her, actually—to put the plates and the numbers aside and take a train ride. He was going to a joint meeting of the American Astronomical Society and the American Association for the Advancement of Science in Washington, DC, and he thought it would be good for her to come along. He even said the observatory would pay for the train ticket. Many of the leading lights in astronomy would be there. "I want you to breeze up to people," he told her.[22]

Probing deep into space, alone, with just her beloved plates was no problem; circulating among earthly beings at a conference filled Cecilia with dread. On arrival, she wandered the hotel conference floor by herself until she caught sight of Ernest Brown. Like Cecilia, Brown had been a student at Cambridge. He had graduated in 1887 with honors in mathematics and went on to become a rower, a mountaineer, a pianist, and an astronomer, whose life's work was spent studying the motion of the moon.

Brown was also a life-long bachelor, which perhaps explains his unique work routine. "He would retire rather early in the evening and as a consequence would awaken usually from three o'clock to five o'clock in the morning. Having fortified himself with a number of cigarettes and a cup of strong coffee from a thermos bottle, he would then set to work in earnest without leaving his bed."[23]

Brown was standing, cigarette in hand, with a small group of men. Cecilia steeled herself. "Driven by despair, I went over and joined them," she wrote to Margaret Harwood, a friend and astronomer at the Maria Mitchell Observatory in Nantucket. "I thought that Brown, of whom I am terrified, would be a good one to start on. . . . [He] encouraged me to join the party . . . and even offered me a cigarette, which I scandalized him by accepting."[24]

How exactly did she "breeze up"? Did she smile and laughingly accept his offer? Did she cup his hand in hers as he put flame to cigarette? There is no record, but Brown must have sensed something; he later "enquired which of the gentlemen I intended to annex for the meeting, and I told him that nothing short of the whole assembly would content me."

After this awkward start, she escaped into astronomy lectures. But Shapley was ever present, and she knew he was watching. "After two days I decided to begin [again] 'breezing up,' as nothing seemed to come my way otherwise," she continued in her letter to Margaret. "I spent my time selecting victims, oscillating from the Physical to the Chemical section in quest of them, leaving astronomy to the other astronomers."

She recognized Karl Taylor Compton. He was a prominent physicist, a summa cum laude graduate of Princeton, who five years later would become president of MIT. From Compton she "gleaned much that was of value. The strain of introducing myself to [him] nearly finished me—but it had to be done."[25] She longed for the plates.

At last the conference drew to a close. "Breezing up" had proven to be far more stressful than digging in to the composition of stars. But Shapley was happy with how his young protégé had performed. In fact, she told Margaret, he came to her rescue. "On the last evening I was worn out, and also a count of my money revealed the fact that I had not enough to buy me a dinner. You can imagine my gratitude when the D.D. asked me to dine with him and Russell, not only for the honour done me, but for the actual food (I was dreadfully hungry)."

When Shapley excused himself early, perhaps by design, Cecilia was left alone with Russell. Different ages, different genders—similar minds. The stress of socializing melted away. After Russell gave a short presentation, "he came and joined me, and we talked the whole evening—about (can you guess?) poetry and ancient Rome. I should not have thought he was the same man, and I feel quite differently about him—certainly I shall not be afraid of him personally any more."[26]

Back at Henrietta's desk, fresh from the Washington trip, Cecilia focused more than ever on applying astrophysics to understanding the spectral lines on the glass plates. The data and her calculations, however, continued their stubborn refusal to cooperate. The pile of cigarette butts grew by the day, but the results did not change. She believed the data, and she trusted her powers of analysis and understanding; but she was also prudent and practical. She was, after all, proposing to contradict existing theories about stellar composition. She followed the traditional scientific process of taking measured steps. She started to publish a series of short papers outlining her results.

She was treading on treacherous ground. The astronomy establishment at the time held a common strong opinion that the composition of all celestial bodies was similar. Especially Eddington; he had staked his entire career on uniformity—that the sun and stars were composed of the same elements as found on earth, with the same relative abundance. He viewed Saha as a marginal figure working out of Calcutta University, generating results that "must be rather shaky."[27]

Cecilia must have been worried about how Eddington would react to her findings, for there was more than science involved. From the very moment when she had transcribed Eddington's talk describing his solar expedition, she had maintained strong feelings for him. When she attended an astronomical conference in Canada in 1924, she confided in a letter to a friend that she had "slipped away from all the people who were trying to think of things to say (and thereby spoiling everything) and went off to stand by myself at the head of the Horseshoe Falls."

"I don't know how long I was there," she continued, "but I seemed to have been there always, when I turned around and found the only other person with whom I should have liked to be there, standing beside me."[28] In her inimitable literary style, she would write later that "it was 15 years before I outgrew my childish dream of playing the Beggar Maid to Eddington's King Cophetua."[29]

Shapley, meanwhile, was hovering. To raise astronomy to the level of a true academic department, Shapley needed Cecilia to get a doctoral degree. The chicken-and-egg problem, of course, was that there was no astronomy department. The only path to a PhD for Cecilia was through the Physics Department. "The redoubtable Chairman of that department was Theodore Lyman," Cecilia remembered, "and Shapley reported to me that he refused to accept a woman candidate."[30]

Lyman had graduated from Harvard the same year that Russell had graduated from Princeton. He served as captain of the Signal Corps in France during World War I. He was so wealthy that when he returned to Harvard to teach physics, he regularly gave his salary back to the university. He traveled the world from Alaska to British East Africa, where he went lion-hunting.[31]

Shapley, however, was more wily than a lion. He was a rapidly rising astronomer, and he was going to get his department. He needed a favor. He composed a letter to Lyman in September 1924. "My dear Lyman," he wrote, "I believe you are the appropriate official to place the approving signature on Miss Payne's candidacy for the doctor of philosophy. She took her preliminary examination in June and is well on her way toward the writing of an exceptionally fine thesis—a monograph on the subject of stellar chemistry."[32]

Shapley then gave Lyman the perfect out; he requested that Lyman send the application to the secretary of Radcliffe College. Lyman's reply came two weeks later. "My dear Shapley: Strictly speaking I have no right to sign Miss Payne's application." But Lyman understood what Shapley was doing—strictly speaking, Cecilia would receive her degree from Radcliffe, not Harvard, which he found himself able to tolerate. "But as I believe that some such action is generally taken in similar cases by the chairmen of other departments, I will gladly affix my signature and forward the paper to the Secretary of Radcliffe."[33]

Cecilia related that she "never knew how Shapley handled the problem."[34] All she knew was that he was pressuring her to collect her papers into a thesis. When she finally complied, Shapley sent her work to Russell for review.

Russell was impressed. He wrote Shapley that he had "eaten it up since I got it yesterday." He closely read the part of her thesis showing that differences in stellar spectra from one star to another were a function of temperature and pressure, not variation in abundances of elements. "I am especially impressed," he continued, "with the wide grasp of the subject, the clarity of the style, and the value of Miss Payne's own results."[35]

The papers were carefully constructed, precise, systematic. But there was nonetheless that conclusion: her analysis of spectral lines, so meticulously preserved for decades on the Harvard Observatory's plates, showed that stars were composed almost entirely of hydrogen and helium. Stars were indeed chemically homogeneous, but their composition did not at all resemble that of the earth's crust. Eddington's uniformity principle was flawed.

To Russell, her conclusion was an insurmountable problem. Even as a young instructor at Princeton from 1910 to 1914, Russell was "fascinated with the apparent similarity of the abundances of elements in the Earth's crust and in the solar atmosphere."[36] Eddington's claim of uniformity throughout the universe was virtually a law, and Russell was a disciple. "The uniformity of nature was a powerful principle accepted by Russell and all the leading astrophysicists of the day," wrote astronomer and historian of science Owen Gingerich. "The earth, with its predominantly iron core; iron meteorites bombarding the earth from outer space; and the overwhelming number of iron lines in the solar spectrum all pointed to the uniformity of nature."[37]

Russell felt compelled to head her off. Ten days after their Washington dinner, he wrote to her: "I am convinced that there is something seriously wrong with the present theory. It is clearly impossible that hydrogen should be a million times more abundant than the metals."[38]

The established men of science: Henry Norris Russell (*left*) and Harlow Shapley

She was cornered now. Shapley's dream of starting an astronomy department at Harvard depended on Cecilia breaking down the barrier and actually earning a doctorate. And Cecilia herself was eager to publish her results in a leading astronomy journal. To do that, however, she would need Russell's blessing. "His word was law," Cecilia wrote later. "If a piece of work received his imprimatur, it could be published; if not, it must be set aside and its author had a hard row to hoe. His word could make or break a young scientist."[39]

On anything non-astronomical, Cecilia would have been happy to go toe-to-toe with Russell. She once wrote to Charlotte Moore Sitterly, a fellow astronomer who had worked for Russell: "I always wanted to challenge him to a reciting competition—I think I could have matched him—in French, German, Latin and Greek poetry, I think I could have beaten him."[40] But "his power in the astronomical world is another matter, and I shall fear that to my dying day, as the fate of such as I could be sealed by him with a word."[41]

One can only imagine the anguish she must have felt. On one hand, her careful study of all those photographic plates after so much preparation—working in the Cavendish Lab, listening to the lectures of Rutherford and Bohr, studying in the library of the Cambridge Observatory—had produced results that were indisputable. On the other, "the winds in physics were blowing against Payne's findings."[42] If she did not accommodate Russell's skepticism of her conclusions, her work would never see the light of day.

And so, instead of defying Eddington, Russell, and indeed the entire astronomical community, she defied herself. "Although hydrogen and helium are manifestly very abundant in stellar atmospheres," she wrote in her thesis, *Stellar Atmospheres*, "the actual values derived from the estimates of marginal appearance are regarded as spurious." Her provocative results that hydrogen dominated the composition of stars, she concluded, were "almost certainly not real."[43]

"Almost certainly." As DeVorkin puts it, "Payne had to make the changes Russell dictated, but she was crafty about it."[44] She had chosen her words carefully—"in a manner that was designed to record for posterity that she was the first to make this observation, right or wrong. In so doing, Payne can be credited with profound political acumen."[45]

Exhausted after two years of nonstop work, Cecilia booked passage on a steamship bound for London. It was her first vacation since arriving in the United States, her first trip back home. It was June 10, 1925. She wrote to Shapley on the ship's stationery. "I left a complete MS of my thesis on the desk with the proof." She went on to make sure that the D. D. knew how grateful she was for his support, and that she would return in a few weeks. "I am not such a fool as not to realise what you have done for me; and it is beyond both hope and belief."[46]

Shapley was ecstatic. He had now presided over Harvard's first doctoral thesis in astronomy. As a result, he got what he wanted: "organizing the degree for Cecilia *de facto* created the Astronomy Department."[47] At first, frugal as always, Shapley worried that he had ordered too many copies of the bound thesis. Would even a hundred astronomers and physicists pay $2.50 for a young scholar's work? He later crowed that the entire edition of six hundred copies had sold out within three years.[48] Russell, too, was pleased. He described Cecilia's thesis as the best he had ever read, except for perhaps that of his former student, Harlow Shapley. And Eddington? He rested easy. His concept of uniformity was still intact. Everyone was happy—except one.

"She always regretted it," says Katherine Haramundanis, Cecilia's daughter, about Cecilia's decision to deny the results of her research. "She knew Russell would not have accepted her thesis if she didn't follow his instruction. She didn't dwell on it. But throughout her life, she lamented that decision."[49]

There have been a number of explanations offered for why Cecilia was warned rather than celebrated. Some believe it was blatant prejudice. "She was bullied," maintained Jesse Greenstein, a Harvard-trained astronomer who knew Cecilia.[50] Certainly gender bias was rampant at the time. A twenty-five-year-old woman graduate student versus the men's club of established astronomers and physicists was hardly a fair fight. Had she

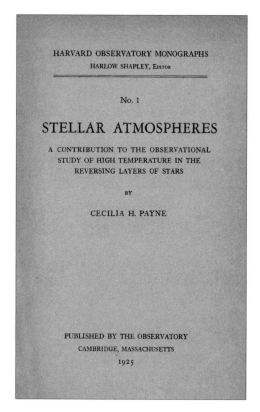

HARVARD OBSERVATORY MONOGRAPHS
HARLOW SHAPLEY, Editor

No. 1

STELLAR ATMOSPHERES

A CONTRIBUTION TO THE OBSERVATIONAL
STUDY OF HIGH TEMPERATURE IN THE
REVERSING LAYERS OF STARS

BY

CECILIA H. PAYNE

PUBLISHED BY THE OBSERVATORY
CAMBRIDGE, MASSACHUSETTS
1925

"The most brilliant thesis ever written in astronomy"

been a man, she probably would have been taken more seriously; there would at least have been some follow up, some immediate testing of her claim, which there was not.

But there were other forces at work as well. Meghnad Saha, the physicist whose equations Cecilia used in her analysis, made assumptions that even he acknowledged were untested. And because of its light atomic weight, hydrogen might be skewing the results somehow.

One could argue that the real explanation is more subtle, more complicated. Rather than bullying her, it is more likely that the men of science at the time did not fully appreciate Cecilia's ability to apply what she had learned at the Cavendish Lab to astronomy. To Shapley, she was more of an employee—the latest in a long line of computers—than the true astrophysicist she had become. To Russell, she was a graduate student who needed to be protected from taking the radical step of challenging Eddington's theory of uniformity.

In hindsight, it is clear that Cecilia had determined what stars are made of, one of the most fundamental discoveries in the science of astronomy. The veteran astronomers of England and Germany and America, however, did not have the combination of rigorous training in atomic theory and access to data that Cecilia did. As a result, they could not peer as deeply into the universe as she could. And they simply could not—would not—admit that.

At least not yet.

17

As the astronomer Margaret Harwood walked down the hall of the Harvard College Observatory in the spring of 1925, she heard something unexpected: the sound of someone crying.

Harwood had graduated Phi Beta Kappa from Radcliffe in 1907. After graduation, she had worked at the Harvard Observatory before being appointed director of the Maria Mitchell Observatory in Nantucket. There was a working relationship between the two observatories, and she often visited Harvard.

Concerned, she listened carefully and then approached the source of obvious distress. She found Cecilia in her office, alone, weeping inconsolably. Cecilia told Margaret that it had been hours since the final oral examination for her PhD, and she had heard nothing yet. She had to have failed.

Harwood immediately went to the director's office. Shapley was completely taken aback. He told Harwood that Cecilia was considered so brilliant that it didn't occur to them to let her know that she had passed.[1]

Many readers of Cecilia Payne's thesis saw it solely as a treatise on the new astrophysics, parsing its charts and tables and calculations for insights into the physics and chemistry of the stars. Others, however, especially in years to come, regarded *Stellar Atmospheres* as an illustration of what a true explorer Cecilia was. Her accomplishment was not like crossing an ocean solo, or planting a flag on the surface of the moon. Rather, the frontier was in her mind. She was probing deep space—every bit as exploratory as any terrestrial expedition—but it was a mental, not a physical feat. Cecilia, typically, saw it more humbly: "All that I have done," she wrote later, "is respond to the quickening influence of the Universe."[2]

As *Stellar Atmospheres* gradually gained recognition, even flinty Ernest Rutherford joined the family of those praising Cecilia, although more as a curmudgeonly old uncle than as a nurturing elder. The Cambridge physics professor confided to Bertha Swirles—an honors student in mathematics and physics at Cambridge, three years younger than Cecilia—that Cecilia "is doing well in astrophysics."[3] There is no record of whether he communicated his opinion directly to Cecilia.

In the first few months after publication, her work did not move scientific thought very much. Resistance to her radical conclusion about the abundance of hydrogen held strong. During her trip home to England, she took some time to visit Cambridge. There, retracing her familiar path from Newnham to Petty Cury, to the Cavendish Lab, to the Cambridge Observatory, she paid a visit to Eddington, now Sir Arthur. In what she described as a "burst of youthful enthusiasm," she excitedly told him that she believed there was far more hydrogen in the stars than any other atom. That conclusion, of course, directly conflicted with Eddington's uniformity theory; to his thinking, there could perhaps be more hydrogen surrounding a star, but not as part of the star's actual composition. "You don't mean *in* the stars," he told her patiently. "You mean *on* the stars."[4]

After a few weeks, Cecilia sailed back to the United States, where, in the other Cambridge, she had more earthly concerns. In her two years at Harvard, she had become thoroughly American. She admitted to aching at times for the English countryside, but home now was the United States. To her, the "land of opportunity" was not just a slogan. She certainly did not miss British class distinctions: "Here was a land where those horrible words, 'our betters' had no legitimate context. It was an inspiring revelation. . . . It supplied one of the strongest reasons for wishing to make this country my home."[5] She had intended to visit for a year, maybe two; she would stay a lifetime.

Her immediate problem was a familiar one—money. With the completion of her thesis, her fellowship funds came to an abrupt end. There was no family legacy to fall back on, no trust fund to tap. For the first time in her life, she needed a job. Robert Grant Aiken, an American astronomer and mathematician, had read her book and was much impressed. He offered her a research fellowship at the Lick Observatory, operated by the University of California at Berkeley, where he was associate director. When Cecilia told Shapley about the offer, he exploded in fury. He said Aiken should have consulted him; he then offered Cecilia the position of technical assistant at the Harvard Observatory.

Shapley knew that Cecilia was both brilliant and destitute, and he suspected that she wanted to stay at Harvard. He must have guessed that he wouldn't have to offer much to entice her to stay. It would have been a good guess. She had casually asked Eddington about working in England and he told her to forget it—that possibility was no better than when she had left, and the opportunity for her to continue working at Harvard "was not to be lightly put aside."[6]

Perhaps her mind was still too focused on her research to deal with personal finance. Perhaps it was just simple inertia—a coast-to-coast move to California coming too close on the heels of a recent transatlantic journey. Whatever the case, "in my innocence," she recalled in her memoir, "I did not ask how much he was going to pay me, or realize how little it would be. Nor, I think, should I have cared very much. I accepted the offer."[7]

Quickly she "found out the difference between a Fellowship and a job. The former pays at the beginning of the month, the latter at the end. Suddenly I found myself without funds." It clearly never entered her mind to borrow money. Instead, she scrimped. She lived on a diet of potatoes and sausages because they were so distasteful that she could only eat small amounts. She had pawned her father's violin; now, to tide herself over to the end of the month, she pawned her jewelry.[8]

Shapley had her pinned. When she had first entered his office, fellowship funds already in hand and therefore independent, he could only suggest tasks for her to do. Now that she was under his employ, he could mandate. And he did just that, returning to the request he had made when she first arrived. The plates that Cecilia had used to unlock the mystery of the stars held vast amounts of data, but there were still no real standards by which to measure the varying magnitudes of stars' luminosity. He directed Cecilia to turn to that project.

Writing her thesis and book had played to her twin strengths: analysis and imagination. There was no imagination required to do Shapley's bidding now. It was stellar drudgery. "Alas for my beloved spectra," Cecilia lamented. "It was hard to leave them, and to turn to the arid field of standard photometry. But such was my devotion to the Director that I did not refuse him, and I embarked on an endless undertaking."[9]

Bright spots were few and far between, but there were some. With a regular paycheck, meager as it was, she was finally able move out of the crowded dorm and into a small apartment on Concord Avenue across the street from the observatory. The clamor of the dorm was gone, but she still had a roommate. To defray expenses, she lived with Frances Wright, a colleague at the observatory. Frances was almost as tall as Cecilia; but where Cecilia was broad-shouldered, Frances was thin and wiry, with a spirit of adventure that Cecilia would come to appreciate and share.

Frances was three years older than Cecilia, but well behind her gifted roommate in the pursuit of astronomical studies. She, too, was subject to Shapley's financial control. "I enjoyed working under the direction of Miss Payne very much," Frances wrote to Shapley, pleading for permanent work. "I should be interested in almost anything relating to astronomy, though I think I am best fitted for work which uses my eyes.

I can use them more quickly than I can compute numbers."[10] Frances knew her lack of math skills lowered her value to Shapley, so she offered him a discount: "I could take care of myself financially if I earned $1000 a year."[11] Shapley saw it as a bargain and hired her.

At Newnham College, where she had a room to herself, Cecilia had gained her first measure of independence. Now at Harvard, with an apartment of her own, she felt liberated. She was no longer a schoolgirl or student; she was a young woman. With a job! Domestic urges, long dormant, began to emerge. She cooked; she sewed; she even entertained. After spending years bringing a scientific project to conclusion, she marveled at the satisfaction of bringing a terrific meal to the table in just a few hours. As she described it, "I often think that the problems of getting a dinner—the tools, the techniques, the timing and the balance—are very like those of planning and executing a piece of research. She who can do the one can do the other—if she will 'intend her mind' on it. Patience, attention to detail, willingness to wait, these are great qualities in a scholar."[12]

She was applying the principle of gradual learning to her personal life. "I had once pictured myself as a rebel against the feminine role, but in this I was wrong," she would write later. "My rebellion was against being thought, and treated, as inferior."[13]

Another bright spot: there was a crack, and just a crack, in the wall of resistance to her hydrogen findings. Cecilia's reputation as an astronomer was growing, both in America and abroad, as more scientists read, and understood, *Stellar Atmospheres*. Albrecht Unsöld, who now worked at the Mount Wilson Observatory in Pasadena, had completed a thorough and detailed paper on quantum theory applied to the composition of stars. When he compared his results with Cecilia's, he found considerable agreement. Unsöld joined other investigators who found evidence that stars had large amounts of hydrogen in their atmospheres, but the situation was still confusing.[14]

As Cecilia's reputation grew, she gained a number of firsts. To Shapley's delight, she was the first PhD in astronomy at Harvard.

Stellar Atmospheres was lauded as a brilliant thesis. (In his 1926 book *The Internal Constitution of the Stars*, Eddington noted that Cecilia's method for determining relative abundances "is not so wild as we might suppose at first.")[15] As she turned twenty-six, she was the youngest astronomer ever to have a star of distinction placed next to her name in J. M. Cattell's reference work on leading scientists, still called *American Men of Science*. She was the first winner of the Cannon Award (named for Annie) for distinguished contributions to astronomy by a woman. The American astronomer Edwin Hubble reportedly described Cecilia, in what he regarded as a compliment, as "the best man at Harvard."[16]

Obstacles, however, pay no mind to accolades; and there were many obstacles. The Harvard Observatory's equipment was no longer the most modern—other observatories were investing in cutting-edge telescopes with larger apertures. Shapley, however, was reluctant to update the observatory; spending large sums for bigger mirrors on a telescope in cloudy Cambridge was hard to justify. And those colleges with better equipment generally had all-male faculties and would not consider hiring a woman. Even Radcliffe did not have a woman on the faculty.

The observatories themselves were no better. They were designed to take the measure of the night skies, so they were often located in faraway, isolated places. The men who ran them, believing it to be improper for women to spend the night in the company of men, were not disposed to hiring female staff members.[17]

Even visiting an observatory could be difficult. When Cecilia expressed a desire to observe through a Harvard-owned telescope at the Boyden station in South Africa, the director thought it was too risky, hinting that he couldn't guarantee her safety. He cabled Shapley that he believed "a lone woman would be in danger from the blacks."[18]

Then there were the astronomical conferences. In the fall of 1924, Eddington was scheduled to speak at a meeting at Yale, a short train ride away. Cecilia wanted to see him, but Shapley never told her about it. As Cecilia wrote to Margaret Harwood, "perhaps Shapley wanted to spare me the pain of knowing it was going on at a meeting which my sex bars me from attending."[19]

As for work in England, that obstacle would never be overcome. Cecilia had "read" physics at Cambridge instead of mathematics, an essential requirement for a job in astronomy in England. And even if she had the appropriate degree, there were other obstacles. The Royal Observatory, for example, was administered by the Admiralty. One of the requirements for the job of chief assistant was that the candidate had to be able to climb a rope. Despite her stature in the international astronomical world, "I should have failed the test," Cecilia exclaimed. "Rope-climbing has never been my strong point."[20]

Even Shapley himself was an obstacle. Not only did he dictate what Cecilia was to work on, but he wanted his money's worth. His fledgling astronomy department would not survive if it didn't place graduates in the astronomical community. He told Cecilia that with her new doctorate, which he had demanded she secure, she must now begin to teach graduate level astronomy courses. It took her away from her research mission, but she gamely prepared a syllabus and took her place at the lectern. All of her talents came together to make her a memorable teacher. She drew on her dramatic ability, honed in the bedtime stories she had conjured up as a girl and the plays she had performed with her brother and sister. In her lectures, she wove together her love of astronomy—she viewed individual stars and their unique spectral lines as personal friends—and her equally strong love of the other branches of science, of history, religion, and the classics. She created lecture tapestries.

"I listened to a course of hers in variable stars," remembered one student, "and what I learned was the English language. If there was ever anyone who speaks it in the way it was intended to be spoken, it was Cecilia and it was beautiful."[21]

"Cecilia could quote (and without error) Gilbert and Sullivan, T. S. Eliot or Latin epigrams," recalled her former student Jesse Greenstein. "She was a quite extraordinary figure; broadly informed in astronomy, of imposing stature and stormy personality, and widely read in current English and American literature and classical music."[22]

"This is where I really learned what 'chain smoking' was," remembered Owen Gingerich, another of Cecilia's students who would become a professor of astronomy and the history of science at Harvard. "A pack of cigarettes and a single match could get her through the entire period."[23]

Shapley got his money's worth. The classes Cecilia taught were not listed in the Harvard course catalog. She did not have the title of instructor or professor, but rather was paid as Shapley's "technical assistant."[24] He could get away with it because the practice was officially sanctioned. The president of Harvard was still Abbott Lawrence Lowell. He was the very picture of the patrician educator. His official oil portrait, painted by no less than John Singer Sargent, depicts him seated on a kind of wooden throne, wearing a black robe with a crimson collar, looking directly at the portraitist, a full moustache waxed to a point at either end.

Lowell had tried to limit Jewish enrollment at Harvard to 15 percent, and he tried to ban black students from living in the freshman dorms. In both instances, the Harvard Board of Overseers overruled him.[25] The board did not overrule him, however, when he decreed in 1928 that women should not receive teaching appointments from the Harvard Corporation. "I had no official status," Cecilia recalled. "I was paid so little that I was ashamed to admit it to my relations in England. They thought I was coining money in a land of millionaires."[26]

Shapley knew he had a good thing going. Cecilia was extraordinarily valuable to him, and to his goal of building a preeminent astronomy department at Harvard; but he couldn't acknowledge it for fear of losing her. He would admit in a letter to the science editor of *The Literary Digest* that Cecilia "is one of the most outstanding astrophysicists of America, of any and all sexes." But in the same letter, he made a request: "Please do not quote me in any way."[27]

Russell, to his credit, was not so possessive, although of course he didn't need to be. When the Canadian astronomer John Stanley Plaskett

asked him to recommend someone for a position at the Dominion Astrophysical Observatory in British Columbia, Russell mentioned Cecilia,
whom he described as "quite the best of the young folks" in astrophysics.[28]
For Plaskett, however, the prospect of hiring Cecilia was a nonstarter:
"There would be difficulty about the observing end of it with a woman
in this isolated place and I think we can hardly consider her."[29]

Cecilia had great respect for Russell, and she maintained a close enough
relationship with him that she felt comfortable pouring out her frustra-
tions. In a 1930 letter, she confessed that working directly for Shapley
after finishing her thesis had been "a very unhappy time . . . ; the chief
reasons have been (a) personal difficulties within the Observatory particularly with Dr. Shapley . . . (b) disappointment because I received
absolutely no recognition, either official or private, from Harvard University or Radcliffe College; I cannot appear in the catalogues; I do give
lectures, but they are not announced in the catalogue, and I am paid for
(I believe) as 'equipment'; certainly I have no official position such as
instructor."[30]

There is no record of Russell's response, but her complaint seemed to
have an effect. Shapley bumped up her salary to all of $2,700 a year. He
said she could take summers off from teaching and devote them to research. He nominated her to the Harvard Faculty Club, and she was
elected as an associate member. But he went too far when he proposed
that Cecilia's course on variable stars be included in the course catalogue,
and that she be listed as the instructor; Lowell and the dean of the faculty would not allow it. Lowell once said to Shapley, who then repeated
it to Cecilia, that "Miss Payne should never have a position in the University as long as he was alive."[31] It was a frustrating time, especially
because her career in astronomy did not seem to be in her hands. And if
there was ever a hands-on person, it was Cecilia. She needed a break.

Adelaide Ames was still very much Cecilia's Heavenly Twin, but she
was even more under Shapley's thumb. She was unavailable. So it was
that Cecilia's ebullient roommate Frances Wright agreed to accompany
her on a road trip during the summer of 1930. Shapley had once suggested

that Cecilia should visit the Lick Observatory in California to inspect the spectrograph there. When she told Shapley that she and Frances were going to take a cross-country drive together to California, Shapley was not happy—he had forgotten all about his suggestion. But she was undeterred.[32]

They could drive there because Cecilia Payne had bought a car! She paid for a black 1930 two-door Model T Ford with the proceeds of a moonlighting lecture course she taught at Wellesley College.[33] It was a somewhat serendipitous purchase. For the Cecilia of the past, it was so out of character; but for the Cecilia to come, it was the first sign: she was going to live a little.

Encouraged by the observatory's night assistant, Frank Bowie, she bought the car before she had even learned to drive. Frank had offered to teach her. He had the credentials. He often boasted about how he had been responsible for the very first stolen car in all of Cambridge. Memorial Drive became her proving ground. "Many a time, when dawn put an end to observing, he and I burned up the road beside the Charles River," Cecilia recalled. "I can still hear his voice at my elbow, urging me on with 'Step on it, Celia!'"[34]

Someone took a photograph of Cecilia and Frances before they set off. Cecilia, tall (five-foot ten) and broad-shouldered, is sitting on the back bumper; Frances, with short curly hair and wire-rimmed glasses, leans against the spare tire. They drove three thousand miles "over roads very different from the superhighways of today. Between Kansas and California there were only dirt roads."[35] They camped out the whole way in a pup tent. They rode mules in the Grand Canyon, Cecilia sporting a floppy sun hat, Frances wearing a bandanna.

At California's Mount Wilson Observatory, Cecilia met "the legendary figures of Western astronomy," including "the distant, forbidding Director, Walter Sydney Adams," and of course "the great Edwin Hubble." The trip was just the tonic needed, although it ended awkwardly. Cecilia wanted to stay overnight at the Lick Observatory, but there was no room available. She and Frances offered to sleep in their trusty tent.

Cecilia and Frances Wright—day one of their trip, 1930

Tents were not allowed on Mount Hamilton. "We left the Lick Observatory unceremoniously."[36]

Meanwhile, there were more cracks in the prevailing view of what stars are made of. Donald Menzel, Cecilia's competitor from Princeton, was now working at the Lick, studying the chromosphere of the sun with the observatory's incomparable collection of solar eclipse spectral plates. When Russell traveled west to visit, Menzel showed him his meticulously detailed work. Menzel later wrote that Russell "became convinced of the correctness of one of my conclusions, that hydrogen was the dominant element of the solar atmosphere."[37]

Russell had listened quietly to his former student. The sun was a star. And what was true for the solar spectrum had to also be true for all stellar spectra. Russell murmured that perhaps it was time for a "reconnaissance of new territory."[38] In fact, by 1928 Russell was beginning to realize that "the hydrogen abundance problem demanded a complete re-evaluation."[39]

Cecilia did not know of Russell's gradual conversion. Despite being diverted by Shapley from further study of stellar spectra to designing

Cecilia in the Grand Canyon

Frances Wright in the Grand Canyon

standards in photometry—"I am a field naturalist, not a surveyor," she lamented—she was as intensely focused as ever.[40] Over and over in correspondence with Shapley, she repeated the same theme: "I begin to ache for work" (1924 letter from Cleveland) . . . "I could wish to be back at work" (1925 letter after her visit to Cambridge) . . . "Please keep some work for me to do" (1927 letter during another trip to England) . . . "You will I hope leave me a New Year gift of some work to do?" (1928 letter written on board a ship crossing the Atlantic).[41] In each instance, Shapley was happy to comply.

One of the ways Cecilia dealt with the stress of getting what she had asked for was by smoking. Eventually it would take its toll on her; but in the early days, it was those around her who felt the effects. "Cecilia was the most prolific chain smoker I ever knew," recalled Dorrit Hoffleit, an astronomy graduate student at Harvard. She once worried that "Cecilia's smoking might have cost me my degree. With a cigarette between her lips she asked a question, of which I could not grasp a single word."[42]

Cecilia smoked for fifty years, trying to quit now and then but never succeeding. Her daughter, Katherine Haramundanis, wrote: "Once she arrived in her office for the day, particularly when working headlong as was her wont, she chain-smoked, lighting one cigarette from the last. Ashes were strewn broadcast, absent-mindedly, and the ashtray overflowed."[43]

Along the way, there had been another crack in the post-thesis years. This one was not in the monolithic wall of Victorian scientific thinking; it was in the no-time-for-that-now wall of Cecilia's nonstop personal drive. It came in the form of one Norbert Wiener.

In the cast of characters who crossed Cecilia's path in life, Norbert was perhaps the most colorful. He was equal parts brilliant and lazy. He was home-schooled by his father, Leo, a professor of Slavic languages at Harvard, until the age of seven. His father then placed him in public school as an underage fourth grader, where Norbert proved to be completely inept at arithmetic. He was yanked back into home schooling. It was not a happy arrangement—whenever he made a math error, he recalled later, "the gentle and loving father was replaced by the avenger of blood."[44]

Norbert got the message: if he got down to work, he could get out of the house. He got down to work. He graduated from Tufts University at fourteen, and he earned a PhD in mathematical philosophy from Harvard at eighteen. He learned to read and speak seventeen languages. A reporter once labeled him "the most remarkable boy in the world."[45]

Norbert was thought to be the youngest college man in the history of the United States, but he was nevertheless an eighteen-year-old kid. He held several odd jobs before finally landing a professorship at MIT in 1919. He would stay there for forty-five years.

He was still brilliant, but now he had traded laziness for absent-mindedness. During his peregrinations around campus, he would stop in on fellow professors to chat, knocking the ashes off his cigar in the blackboard's chalk tray. He was known to conclude the conversation by asking his colleague which way he had come before stopping in. When he got the answer, he would reply, "Good! That means I've had lunch."[46]

Beneath the quirkiness, however, was a mathematically gifted mind. He devised a probabilistic description of Brownian motion and built a system that improved the accuracy of anti-aircraft guns. His book *Cybernetics* connected biological and electromechanical systems—ranging from telephone networks to the nervous system—using principles of feedback and control. (Today's ubiquitous use of "cyber" can be traced back to him.) He died in 1964, just two months after traveling to the White House to receive one of the first National Medals of Science.

In 1925, all that was to come when thirty-one-year-old Norbert boarded a steamship bound for London. He had just broken up with his longtime girlfriend, Margaret Engemann, a German woman who had recently emigrated with her family to the United States. Thus Norbert was on the rebound when at some point during that multiday voyage he met the young astronomer Cecilia Payne, who had just received her doctorate.[47] He may not have felt love at first sight, but there is no doubt he was smitten.

"I am in steady correspondence with Miss Payne," he wrote to his sister, Constance, that summer, "and I intend to pay her attention in the fall. She is a very fine young woman, well read outside her subject, cultural, straight-forward, and with a lively sense of humor." He did, however, notice that "she is a bit socially uninformed, and has very little idea how to dress." He said that he had worked hard to get to know her and that he thought it was having an effect. "She has a scientific book coming out. . . . I like her, and I think she likes me."[48]

Norbert Wiener

He wrote to his brother, Fritz, around the same time that his new-found flame was "jolly, enjoys things keenly, and is not a bit of a prig." But he also felt his competition, clearly recognizing the undercurrent of ambition in the woman he had come to desire. "She is so devoted to her science that I do not know whether I would stand a chance with her."[49]

Norbert soon felt confident enough to write to his mother and father. Like his earlier letters to his siblings, this one featured the cribbed handwriting of an intense, excitable young man falling madly in love. He excused the hurried scrawling in typical Norbert fashion. "Dear parents: I have mislaid my fountain pen and hence must write in pencil." He related that he had had a good time at the theater with Cecilia: "She has a keen sense of humor and is good company, but has gone through the one-sided development characteristic of the English scholarly woman, accentuated by a measure of poverty."[50]

As the summer heated up, so too did Norbert's passion. He met Cecilia's mother, Emma. He and Cecilia took long walks in London parks. "We are going to take in all the shows worthwhile in London together," he wrote Constance. "Ain't we got fun?"[51] But in another letter, he expressed hope and doubt in equal measures: "I don't know whether a girl with such a promising career ahead of her would think of marriage, but if she would, I suppose a scientific man would stand the best chance."[52]

By summer's end, Norbert was completely captivated. He saw in Cecilia everything a "scientific man" could ever want. He saw them living together in Boston. In August, he wrote Constance that he found Cecilia "awfully nice, awfully cordial, and now that she is rested, really nice looking. (She bobbed her hair by the way.)" Like a charged-up exuberant puppy, he could barely contain himself. "I don't promise you not to be engaged by the time I return home."[53] To his mother, he was a bit more coy. "As you know, we are going back on the same boat. A lot can happen on a boat."[54]

Apparently, not a lot happened. Norbert was an irrepressible presence, but the more he pressed, the more Cecilia backed away. She was living in America now, far from the British view that professional wives "should be pleasing, supportive individuals [who] . . . might ideally help their husbands and share their interests, but no more than this."[55] The role of a professor's wife was not one Cecilia imagined herself playing. She did not want to be Mrs. Norbert Wiener; she wanted to be Cecilia Payne. She did not want to be just a wife; she wanted a career as a scientist. And she certainly did not want to languish in the shadow of someone else's career.

"When I left Miss Payne on the boat, it was with permission to write to her and with the promise of letters from her," Norbert told Constance. But he had been disappointed. "Now, although I have availed myself freely of my permission, the promise has up to the present been barren of results."[56] He wrote to his brother that he had met a friend of theirs who had brought "mail for me from home, but—sad to relate!—no letter from Miss Payne. Alas! I have no status!"[57]

Alas indeed. Norbert finally realized that it had only been his flame all along. He moved on. Or, more accurately, moved back. In the spring of the following year, he wrote to his brother. Whereas his previous letters had been handwritten, the hastily scrawled words always lagging behind the pace of the emotion expressed, this letter was typed, as cold between the lines as the others had been hot. "As for me, you know of my coming marriage with Marguerite. I had a fine visit to her recently, and the marriage comes in three weeks."[58] Norbert had patched it up with Margaret Engemann, a woman who "vowed to become the caretaker and protector of her high-strung husband as he careened through his accelerating career."[59]

They were two headstrong people, Cecilia and Norbert, so similar in their need to observe, to analyze, to understand. Too similar? One can only contemplate what might have been.

Cecilia's personal wall was preserved, but the scientific one—the wall of belief that earthly metals were the principal elements of stars—was finally coming apart. Henry Norris Russell, the man who had championed the idea that the entire universe—stars, planets, asteroids, interstellar dust—was uniform in composition, was at last concluding otherwise. Reluctantly, for sure, but conclusively.

As Russell edged closer to what Cecilia had discovered, so too did the rest of the astronomical community, for several reasons. For one thing, Russell's path to discovery was different from hers. Cecilia had applied her knowledge of astrophysics to starlight captured on photographic plates. Russell, on the other hand, presented an entirely independent argument based on the physics of the hydrogen atom itself, not on the spectra of stars. But he got to the same place. "The obvious explanation," he concluded, "that hydrogen is far more abundant than the other elements—appears to be the only one."[60]

Thus it was four years after Cecilia's book was published that Russell finally changed his mind and wrote the paper that persuaded the astronomical community that high abundance of hydrogen is what characterizes the universe. "The outer portions of [giant] stars must be almost pure hydrogen, with hardly more than a smell of metallic vapors in it," he wrote in 1929.[61]

What Russell had determined, in his own way, was what we know today: roughly 98 percent of the mass of our sun (and thus of all stars) is made up of hydrogen and helium. Hydrogen dominates by far—for every 2,000 hydrogen atoms there are only 126 helium atoms, and only one atom each of the next-most abundant elements, oxygen and carbon.

The dam had burst. If Henry Norris Russell proclaimed it, it had to be true. The entire astronomical community soon rushed to concur. In 1933, Robert d'Escourt Atkinson, a British physicist and astronomical clockmaster, summed up the new consensus: "Russell has recently shown that the percentage of hydrogen in stars is probably very much greater even at the present time than has generally been supposed. . . . It seems very reasonable to assume that in its initial state any star, or indeed the entire universe, was composed solely of hydrogen."[62]

Russell compared his results with Cecilia's calculations and found what he considered to be "very gratifying agreement especially when it is considered that Miss Payne's results were determined by a different theoretical method, with instruments of a quite different type (Harvard objective prisms)."[63]

A nod to Cecilia, but David DeVorkin contends that "he was somewhat less than willing to indicate to his readership that he had made a significant reversal."[64] So he buried it toward the end of his paper. He never admitted that he was the one who convinced Cecilia to characterize her results as "almost certainly not real."

So, "did Russell cheat Cecilia Payne of an epoch-making discovery?" asks her former student Owen Gingerich.[65] It's not an easy question to answer. New ideas were emerging fast, with little time for reflection, much less verification. The model of the atom developed by Rutherford and Bohr was still new. Also, writes DeVorkin, "she certainly did not convince Russell, and if Russell knew anything, he knew that her argument would not convince others. . . . He knew what the community would accept."[66]

What the community would accept. At the time, Russell was the one who received credit for discovering what stars were made of. He had what Cecilia did not. She was a young woman with a newly minted PhD. He had connections, stature, prestige, in short, authority—all the essential characteristics required to move an entire community of established thought to a radical new place.

As Owen Gingerich perceptively pointed out, "Like Moses, Cecilia had made a truly memorable contribution. And like Moses, she had glimpsed the promised land, but hadn't quite got there. With what we know today, we could wish that it were otherwise, but . . . it is the person who persuades his colleagues of a new result who gets the credit."[67]

For Cecilia, as left-handed as it was, Russell's acknowledgment was sweet vindication. And it further strengthened her reputation in the astronomical world—although it is proverbially true that one cannot pay the rent with reputation. Despite her new-found reputation, Cecilia still faced obstacles because of her gender. When Russell was looking to hire someone to groom as his successor, he remarked, obviously thinking of Payne, that the best candidate "alas, is a woman!"[68]

Meanwhile, Shapley was a busy man, energetically tending to his legacy. Cecilia had earned the first PhD at the Harvard Observatory, but technically it was a Radcliffe degree. Shapley was still determined to found the *Harvard* Department of Astronomy. He cajoled, he demanded, he inveigled. He knew he had it when in 1928 he finally got permission to hire a department chair. "I could have done it," Cecilia said to herself. "Who knew the ropes better?" But Lawrence Lowell was still president. "It was 'impossible'; the University would never permit it."[69]

Shapley had to know his star researcher was more than qualified, but he was not inclined to go head-to-head with Lowell. He called her into his office and asked, "How much would it disturb you if Harry Plaskett were to come to the Harvard Observatory?"

Harry Plaskett certainly had the pedigree. He had worked at the Dominion Astrophysical Observatory in British Columbia and was an expert in spectrophotometry. Cecilia braced for trouble—years earlier, when she was a graduate student, she had very publicly differed with Harry's father, astronomer J. S. Plaskett, during an international conference. Also, she knew Shapley; he undoubtedly would warn Plaskett that she could be difficult. So, on the day he arrived, she "dressed with extra attention and put a blue ribbon in my hair."

"You're not at all like I expected," exclaimed Harry. "What *had* he expected?" Cecilia wondered. They eventually became good friends, but they never discussed astronomy. "He treated me as a woman, not as a scientist," she wrote later.[70]

It was not the first time that Cecilia would think about discrimination. In her early years at Harvard, when astronomers visited, when she was so deep in late-night conversation with them that they closed down Harvard Square coffeehouses, there was a sense of scholarly equals. "In that heady atmosphere," she recalled, "a woman did not degenerate into the abominable stereotype of the *Femme savante*, that combination of conscious erudition and affected coyness that suggests 'It's really not *womanly* to know as much as I do.'"[71]

With the department up and functioning with a newly installed chair, and with graduate students flowing in, what Plaskett expected, and Shapley too, was that Cecilia would continue to teach. And so she did, over the years becoming one of the most popular instructors at the observatory. Later she would write a textbook, *Introduction to Astronomy*.[72] It was not like most undergraduate textbooks. To Cecilia, the line between art and science was not just blurry; it was porous. Art flowed into science, and science into art. "Good scientific thought has an esthetic perfection," she once observed.[73] So it was that each of the fifteen chapters began with a literary quotation. A quote from Goethe's *Faust* for the chapter on "The Earth," from Shakespeare's *Love's Labour's Lost* for "The Sun," and from Gerard Manley Hopkins for "Stars," clearly her favorite topic:

Look at the stars! look, look up at the skies!
O look at all the fire-folk sitting in the air![74]

Each semester, there was a waiting list to take her introductory course. It was such a treat to sit in the observatory's amphitheater, to listen, to take notes. Her lecture style was like no other. When her former students were asked to describe it, they would all use the same words. "She rarely looked at the audience. Instead her sight was turned skyward in the direction of the stars. It was as if she were conversing with them while the audience eavesdropped on the one-way conversation."[75]

The lecturer was riveting—and unlisted. Students could find "Introductory Astronomy" in the course catalog, but not the name of the "professor." But then, there were *no* women professors at Harvard. There never had been. "My nameless status remained nameless," recalled Cecilia.[76]

Plaskett did not last long. When Oxford called in 1932 with an offer of the position of Savilian Professor of Astronomy, he took it. Cecilia, British by birth and with a growing list of awards and honors and responsibilities, considered herself to be just as qualified for the position. She had to admit to herself that she was jealous. "Not for the first time, I felt I had been passed over because I was a woman."[77]

Chagrined, Shapley now had to find another department chair. He thought the problem was that the job needed to be defined better. "What this Observatory needs is a spectroscopist!" he exclaimed. "I am a spectroscopist!" Cecilia replied. Her retort was indignant; but the listener's ears were deaf. Shapley first offered the position to Otto Struve, head of the Yerkes Observatory. Years later Struve told Cecilia that Shapley assured him that "Miss Payne shall give up spectroscopy—you will have a free hand."[78]

When Struve turned him down, Shapley again called Cecilia into his office. His words were familiar: "How much would it disturb you if Donald Menzel were to come to the Observatory?"[79] Shapley had uttered

Donald Menzel

the words "Menzel has come!" years before, when Menzel was a graduate student working on the same topic. Now he was coming to be a staff astronomer and assistant professor at Harvard. After Menzel's arrival, he and Cecilia existed in a state of "armed truce." "This was a grave loss to me," she wrote, "and perhaps to science too."[80]

Only later did Cecilia realize that Shapley was guided by a divide-and-rule system. He made sure that the staff worked in silos with little communication, much less teamwork. This system meant there was no threat to his command; everyone depended on him.

Cecilia was working steadily now. She had an apartment, albeit with a roommate; she had a paycheck, albeit paltry. She had terrestrial students and stellar friends around her. Her eyes were on the stars, not on the day-to-day world. The stock market crash of 1929 meant nothing—she lived frugally, and her income did not permit investing. The very uneventfulness of it all allowed her to look forward to a life devoted solely to scientific work.

But the steadiness and sweetness were marred with sadness. In May of 1933 she received the news that her close friend Betty Leaf was dead. She had seen Betty on trips home, pulled the snapshot of her from her wallet when she needed comfort. Betty was like a family member, and now the picture was all that was left. The details were murky; all Cecilia knew was that Betty had drowned. And then there was another tragedy.

An article in the *Boston Herald* on June 28, 1932, opened: "Airplane pilots, police and summer residents today conducted an unsuccessful search of Squam Lake for the body of Miss Adelaide Ames, Harvard Observatory astronomer, drowned yesterday as she was canoeing with a companion."[81]

Adelaide had been out on the lake one afternoon with a friend when a squall struck. The two of them were tossed out of the canoe into the water. Although a good swimmer, Adelaide soon disappeared from the surface. Shapley himself rushed to the lake and took charge of the search. It would take ten days to recover the body.

Cecilia was devastated. Her closest friend from college and the other half of the Heavenly Twins—gone forever. She had described Betty and Adelaide both in such a similar way. Betty: "At Newnham she and I had been inseparable. I can hardly bear to speak of her." Adelaide: "In my first year at Harvard we had been inseparable. . . . In later years there had been other friends, but none whom I loved as I loved her."[82]

Cecilia felt the acute guilt of the survivor. One drowning would have been tragedy enough, but two? "Adelaide and Betty—all that I was not, beautiful, delicate, beloved—were dead and I was alive. . . . I was absorbed in my work, shy and unattractive. What was I giving? I made a silent resolve: I would open my heart to the world, I would embrace life."[83]

The twin tragedies marked a turning point in Cecilia's life. She woke up from the trance of work she had been in since her arrival in the United States. Instead of focusing on the stars, she would focus on the here and now, focus on earthly friends, "embrace life." And when she came out of the trance, she "embraced life" the same way she did everything else: with focus and determination. Cecilia Payne fell in love.

There is no historical record as to who was the subject of her first-time affection. As Cecilia described it later, however, she was acting "unreasonably, groundlessly, but nonetheless thoroughly (for I am nothing if not thorough)." Whoever it was, he found himself in the cross-hairs of an intense thirty-three-year-old woman. He backed away. "I fell into a state of despair," wrote Cecilia, "that seems to me now to have been comically Victorian. I felt that my life had been shattered, that I must make a break with the past."[84]

To break with the past, she had to travel, leave the United States for a while, get away from Harvard, from the observatory. Many people in her situation would flee to a warm clime or a romantic locale. Cecilia's getaway-from-it-all destination, however, was not Paris or the Riviera or London. It would be Pulkovo—a remote observatory on the outskirts of St. Petersburg, where spoken words were the whispered preparations for war, where firewood was scarce, food scarcer. Where drama was in the air—fitting, for Cecilia's life was about to take another dramatic turn.

18

Sergei Gaposchkin was scrappy. He always had been.

He was born in Crimea in 1898, one of nine children of the day laborer Illarion Gaposchkin and his wife Ekaterina. There was barely enough money for food. Sergei left school in 1915 and then worked in a textile factory in Moscow until he was called up for army service in 1917. His military career, though, was short-lived—it ended with the collapse of the Russian Empire and the outbreak of war. He then walked for several months back to Crimea to turn in his rifle.[1]

Hardship followed him. He worked as a police officer by day and attended school at night. In the spring of 1920, with civil war raging, both his parents and several siblings died of typhus. In October of that year, he signed on as a mate on a steamship transporting flour. The ship promptly sailed into a huge storm in the Sea of Azov off the coast of Ukraine. The crew managed to get to Constantinople, only to be trapped there when the Bolsheviks defeated the White Army.

With no funds and no papers, he scrounged for odd jobs until Russian émigré friends helped him travel to Bulgaria, where he attended college. From there he continued on to Berlin. He enrolled at the Institute for Foreigners to learn German and to prepare for applying to the local university. It took him almost a decade, but since childhood he had been fascinated by the stars, and in 1932 he got what he wanted: a PhD in astronomy.

Berlin was a vibrant and cultural city, home to a host of accomplished scientists. Among Sergei's professors were Albert Einstein and the physicist Max Planck. Paul Guthnick, director of the Babelsberg Observatory on the outskirts of Berlin, was impressed by Sergei's obvious drive and offered his ambitious student an assistant's job at the observatory, where Sergei had worked as a volunteer while in school. By the summer of 1933, however, Sergei was in yet another tight spot: he was a Russian in Germany just as Adolf Hitler was assuming power.

Realizing that he could be arrested and sent to a concentration camp, Sergei tried to live quietly and keep his head down. Guthnick, however, was feeling the heat. The National Socialists were putting more and more pressure on him to turn in his Russian employee. Fearing for his own position, Guthnick called Sergei into his office and told him that he had to leave. But where could he go? He could not stay in Germany; he could not return to Russia.

Sergei knew that in four days there would be an astronomy conference in Göttingen, about two hundred miles away. He had heard that a number of foreign astronomers would be attending the meeting. He may have had little money and no passport, but he did have a bicycle. If he pedaled hard, he could make it. He packed the bike with four days' provisions and set off.

As turmoil swirled around Sergei, Cecilia was nearing Leningrad. Her trip had begun with a tour of observatories in northern Europe: the Cambridge Observatory, of course, as well as observatories in Leiden, Copenhagen, and Stockholm. At the Lund Observatory in Stockholm, she "did overdo in potations of Swedish Punch," she admitted later. "I learned the lesson that one should know one's limitations." The tragic deaths of Betty and Adelaide were still fresh in her mind. She needed stronger tonic.

"Every one of these visits was pleasant, social and superficial. But I was bent on a more serious trip, the visit to Pulkova."[2]

Pulkovo (to use today's spelling) was considered the astronomical capital of the world in the mid-nineteenth century.[3] The observatory's 15-inch refractor was the twin of the 15-inch at Harvard, and there was close contact between the two observatories. Cecilia had been introduced to the director, Boris Gerasimovič, at a meeting of the International Astronomical Union. He invited her to visit, and she agreed to come.[4]

Cecilia's journey to Pulkovo was by land, sea, and air—none of them easy. First she took a ship from Sweden to Finland. Her fluency in English, French, German, and Italian was of little help after she arrived; she resolved to learn Finnish when she returned to Harvard.[5]

From Helsinki, it was a quick twenty-minute flight by hydroplane to Reval (now Tallinn), in Estonia. As she waited for the plane, curious as always, she tiptoed down the gangplank toward the sea. It was low tide, and the wooden boards were covered with green slime. Thus began a slow-motion, but unstoppable, bit of comedy. "My foot slipped, and I slid very slowly but inexorably into the Baltic. As I stood helpless, waist deep in the water, two airport policemen came rushing to the rescue, waving little white towels, pulled me out and rubbed me down vigorously to dry me off. I was bundled, still dripping, into the plane."

In Reval, Cecilia was met by the astronomer Ernst Öpik. Öpik's specialty was the study of so-called minor bodies—comets, meteors, asteroids. In 1931, a happier time, he and Shapley had led a meteor expedition in Arizona. But on this day in 1933, Estonia was on high alert; when he heard of Cecilia's intentions to go to Russia, Öpik insisted that she cancel her trip because it was too dangerous. As Cecilia pushed back, he pointed out that the Russian visa on her passport had expired. He looked at her triumphantly, believing he had won the argument. He should have known better.

The next day Cecilia went to the American consul in Reval seeking advice. The consul informed her that the United States had no diplomatic ties with Russia; if she got into trouble, they would not be able to help. He advised her not to go. He should have known better.

She went to the Russian embassy. She felt like she had the lead in a "rather grim Gilbertian opera. I passed from one gorgeous office to another, and finally had the satisfaction of seeing my visa updated, with apologies for the 'mistake.'" She bought a train ticket for Leningrad.

Cecilia described the journey to Leningrad (now St. Petersburg) as "weird"; for most people, it would have been frightening. She was alone on the train as it crossed into Russia. The border guards insisted on counting her money—a little embarrassing as she carried all her currency in a money belt underneath her clothes. She experienced a brief moment of panic when stepping off the train in Leningrad—it was a cold, bleak station with no porters, and she could not read the signs.

But then she spotted her host. Boris Gerasimovič had just taken over as director of the Pulkovo Observatory. Dressed in a suit and sweater, with a beaming, kindly face and piercing eyes, he emerged out of the drab grayness, a "blessed sight," as Cecilia recalled. He had come in a small pickup with a driver. Because it was illegal for three people to sit in the front seat, Cecilia rode to Pulkovo sitting on the bed of the truck.

She spent two weeks at Pulkovo but felt as though she had "experienced a lifetime." Food was severely rationed. She had brought coffee, and the staff held a party to celebrate because no one had tasted coffee for years. One night carrots were served as a special treat. Gerasimovič, director of one of the great observatories of the world, admitted that he had stolen them from a neighbor's garden. Cecilia had trouble eating the food. Not only was it unappetizing, but she worried that she was reducing their already limited portions. Everyone was afraid. One young woman staffer begged Cecilia to help her escape. Cecilia was appalled; there was nothing she could do.

When she toured the observatory, she marveled at the Great Telescope. She also took note of the astrograph that Gerasimovič had obtained in order to take rudimentary stellar photographs. Careful, she was warned, "there's a bee's nest behind the shutter. How long, I asked, had it been there? *Seven years!*"

Gerasimovič was thrilled that Cecilia had made the effort to come to his beleaguered observatory. "An American angel came down from the sky in the person of [Cecilia]," he wrote to Shapley in late August 1933. "She left here a bit of typical Harvard atmosphere—a real inspiration for me."[6]

If the stop at Pulkovo was meant to shake thoughts of her own personal tragedies from her mind, it worked. On the drive back to the Leningrad train station, Cecilia noted the troops and military planes along the way. "We are preparing," Gerasimovič replied to her questioning look, "against our enemy." When Cecilia asked what enemy that was, he replied, "Germany."

She sensed that she would not see Gerasimovič again and felt sad when he gave her an embroidered tablecloth and explained that "it is a custom with us, when a friend is going on a journey, to give him a tablecloth." Four years later, Gerasimovič would be pulled from a train returning from Moscow to Leningrad by his own countrymen. He was accused of various crimes, including not complying with Marxist-Leninist ideology because he published scientific papers in non-Soviet journals. He was executed in Leningrad on November 30, 1937.[7]

On the train to Berlin, Cecilia debated whether to go on to an astronomical conference in Göttingen—she had gotten violently sick in Reval after eating a roasted chicken, the first meat she had tasted in two weeks. "But some good angel prompted me to take courage," she wrote later. She pressed on.

Established in 1863, the German Astronomical Society (Astronomische Gesellschaft) was the second oldest astronomical organization after the Royal Astronomical Society. The 1933 annual meeting was in Göttingen, a picturesque university town in Lower Saxony. The tension and anxiety that Cecilia found in Leningrad was in the air here, as well. To her relief and joy, Arthur Eddington was at the meeting. But he was busy chatting with important colleagues. Cecilia took a seat toward the back of the large University of Göttingen auditorium.

"Miss Payne? Miss Payne?" A conference official called out her name, a box of mail in his arms. "Here!" she answered. A young man sitting nearby glanced up, a look of surprise on his face. "Sind Sie Miss Payne?" he asked.

Cecilia said she was indeed Miss Payne. The man introduced himself as Sergei Gaposchkin. He was obviously taken aback—given Cecilia's reputation, he was unprepared to see the young woman in front of him. "He had expected, as I learned afterwards, that Miss Payne would be a little old lady," Cecilia recalled later, "and was surprised to find her no older than himself."

Sergei had cycled for four days to the conference, "in the hope of meeting me," wrote Cecilia later. He may have been surprised that she was a young woman, only two years younger than he was, but he was on a mission. He thrust into her hand an elaborate history of his life, written in German. He asked her to read it, and to help him.

When Cecilia returned to her room later that day, she read Sergei's account. She read of his birth, his siblings, his multitude of jobs, his military experience, his "material distress," his "nearly insurmountable difficulties."[8] She read it again—and again.

She saw someone much like herself—someone who had resolved, just as she had, to become an astronomer. Someone who had achieved against great odds. She had come a long way in a few short weeks. She had started this trip naively unaware of what was behind the expression "Heil Hitler," remarking later that she was "so innocent . . . of what was happening in the world." She was ending the trip confronted by a man seeking her help against Nazi persecution. "I had not spent many sleepless nights," she wrote later, "but that one was sleepless. Perhaps this, I thought, is my one chance to do something for someone who needs and deserves it."

The next morning the conference was not at the top of her mind. She spoke with several astronomers about Sergei and his plight. They all said that they could recommend him and that he was a good astronomer, but he could not stay in Germany. She began to understand: "He had been born in Russia; that was enough. (Who is your enemy? I had asked Gerasimovič; the picture fitted together.)"[9]

This time it was Cecilia who sought out Sergei. They left the auditorium and walked together up the main street of Göttingen. Sergei kept looking around to make sure they were not being overheard. She was honest. She told him she could make no promises, but she would do what she could. Sergei listened, "nervously giving the salute to passing 'brown shirts.'"[10]

When she left Göttingen, Cecilia headed straight for England. On August 12, 1933, on the train from Cambridge to London, she took out paper and a pen and started writing a long letter to Shapley.[11] The letter was classic Cecilia—detailed, thorough, organized. But it was also something else, something that up to now had not characterized the correspondence between her and Shapley. It was emotional.

She had told Sergei to write to Shapley. Shapley had acknowledged receiving Sergei's letter in an earlier note to Cecilia. (Shapley had also been appealed to on Sergei's behalf by Guthnick but had replied that there was nothing he could do for the Russian.)[12] Cecilia was taking no chances. She summarized Sergei's difficult circuitous route to getting his degree. She explained that Sergei had to leave Germany if he wished to continue working in astronomy. Whereas she had fled England on a steamship, "he came from Berlin to Göttingen on his bicycle (four days' journey) carrying his provisions with him."

On page three she started a new thought. "So far I have set forth the facts as dispassionately as I can." Informing Shapley that she now was going to "speak my own mind frankly," she came to the point: "I think we should find a place for him in America, probably at Harvard."

Like an attorney arguing before the bench, she laid out the obstacles and how to deal with them. Sergei was without nationality, but she was talking with "influential friends" in London about getting him an international passport. She judged his work to date as "good but not brilliant." Finally, on the matter of how and what to pay him, she appealed to

Shapley's frugality: "He is willing to work for very little." She then skipped right to the negotiation. "I anticipate that a thousand dollars is a maximum—eight hundred would do."

Shapley was a smart man. He was smart enough to know that in Cecilia he had a brilliant, and inexpensive, worker who did his bidding with little—or at least with an acceptable level of—complaint. He could not afford to lose her. So when she wrote, "forgive me if I speak strongly . . . it is not a rash impulse, and it comes from the heart," he must have realized that he needed to pay attention. He was also smart enough to be able to read between the lines: "He is small but strong, and of a happy (!) and simple disposition." Some lines, though, needed no interpretation: "I cannot personally see any other course than to help him."

Cecilia arrived back at Harvard in September 1933. Shapley was non-committal at first. He wrote Sergei: "I have your letter and I have talked over your situation with Miss Payne, who has just returned from Europe. Whether anything can be done to help with the progress of your scientific work I am unable to say at the present time."[13]

"I had never tried to exert influence before," Cecilia remembered later, "but I tried it now."[14] She took it upon herself to travel to Washington to push along a visa for a stateless man. She pressured Shapley. It only took a month for him to give in. "Dear Dr. Gaposchkin," he wrote to Sergei on October 10, "I am glad to offer you a position at the Harvard College Observatory."[15]

Shapley also wrote to the American consul general in Berlin. "As Director of the Harvard Observatory I have invited Dr. Sergei Illarionowitsch Gaposchkin to join our staff as research assistant. This is a full-time position and it will be ready for Dr. Gaposchkin's occupancy as soon as he can arrange to come to Harvard." Shapley ended the letter with a flourish: "It will be a personal favor to me and a distinct scientific service to Harvard and American astronomy if you can help Dr. Gaposchkin in his arrangements to come to America."[16]

Shapley's letters were Sergei's ticket out of Nazi Germany; the visa Cecilia expedited was his ticket into the United States. Three weeks later, it was done. In November 1933, Sergei penned a grateful note to Shapley: "I got my visa and I go to America from Liverpool on 18th November and the steamer 'Georgie' arrives at Boston on 25th November."[17]

By December, Sergei had his own desk at the Harvard Observatory. He immediately joined Cecilia in trying to understand the Cepheids, a type of variable star whose periods of variation from bright to dim are correlated with their luminosity. But there were other forces at work besides trying to explain mysterious stars. Timing is everything. Both Cecilia and Sergei were ready, finally, to settle down. Professionally, Cecilia found Sergei to be as devoted to astronomy as she was; personally, she found him to be delightful. A weightlifter and body builder, Sergei would demonstrate on the banks of the Charles River how he could hold a headstand for a long time, to a laughing audience of one.

He had benefited greatly from Cecilia's efforts, but his appointment at Harvard was nonetheless temporary. The advantage of linking himself to an American—Cecilia had become an American citizen in 1931—was all too obvious. And Cecilia knew that if they married, she would not be just an astronomer's wife; Sergei would depend on her financially, guaranteeing that she would continue working.

Just as the spiriting of Sergei out of Europe was quick, so too was the next life-changing event. Three months after Sergei stepped onto American soil, he and Cecilia, without a word to anyone besides Shapley, slipped out of the observatory and headed to City Hall in New York. The death of her two friends, the politics of Russia and Germany, a man's desperate plea for help, it all "led to the uniting of two lives, the flowing of two rivers, bound for the same goal, into one channel," she wrote later. "In March 1934 I became Cecilia Payne-Gaposchkin."[18]

When he heard the news of the impending wedding, Shapley, typically, sprang into action. He arranged for friends of his in New York to throw a champagne party for the newlyweds. In a letter written a day after the wedding, Cecilia told Shapley that his friends "chaperoned me, and entertained us on our return from City Hall, in an unnecessarily regal

Cecilia and Sergei after their wedding, 1934

fashion." She also wrote that she was "deeply touched by the roses, and I wore them on Monday when we were married."[19]

Shapley calmed Emma down. Cecilia's mother was concerned about her daughter's hasty marriage to a Russian exile. After receiving Shapley's assurances, Emma replied that she was relieved that Cecilia had found a "congenial mate."[20] And he soothed Frances Wright. Frances was wounded—she was going to have to move out of the apartment she shared with Cecilia so that Sergei could move in. Cecilia thanked him, writing, "I could see that when I left she was quite stunned—but she cannot have been really surprised, I suppose."[21]

Cecilia was in a dream. "I had never thought that such happiness could be for me," she wrote Shapley. The astronomical community, however, was astonished. One of its own, a thirty-four-year-old English woman, had up and married a Russian émigré fresh off the boat. To the close-knit Harvard Observatory family, Sergei and Cecilia's sudden disappearance was head-spinning. One story making the rounds—undoubtedly a myth, but revealing in its believability—was that upon hearing that Cecilia had "eloped," Annie Cannon had fainted dead away. It was Henry Norris Russell, though, who best captured the mood of Cecilia's colleagues. In a letter to the MIT physicist Joseph Boyce, Russell commented, "I sincerely hope that it turns out splendidly, but I keep wondering how it happened."[22]

19

The honeymoon was not a trip to an idyllic spot in the Caribbean or even Europe, but it was nonetheless romantic—astronomy-style romantic. Cecilia pulled out the map, and together she and Sergei retraced the route she and Frances had followed in that old Model T four years earlier. With Cecilia doing the introductions, Sergei met the great American astronomers of the West and toured their observatories. From the Flagstaff Observatory to Mount Wilson to the California Institute of Technology in Pasadena, they moved easily among the astronomical community, dining out and basking in the dry southwestern heat and the attention of being newlyweds. They returned east on the SS *President Pierce* via the Panama Canal. Cecilia had left the pup tent in Cambridge.

Back at the Harvard Observatory, they dove into their study of variable stars with renewed energy. Cecilia greatly missed studying stellar spectra, but the Dear Director's directive could not be ignored. It was not easy. Although variable stars were intellectually interesting—their luminosities varied from dim to bright in mysterious ways—they had long been the province of amateur astronomers and thus bore the ultimate astronomical pejorative: Cecilia's thesis competitor, Donald Menzel, had once declared that "variables were for amateurs, not professionals."[1] Real astronomers, that is, did not waste their time on them.

Shapley didn't make it any easier; he had been known to describe variables as "pathological stars." But if he assigned variables to his new team, so be it. At one of his weekly free-for-all "Harlow Square" meetings, while Shapley was in the middle of extolling the virtues of the observatory's voluminous collection of plates, Cecilia interrupted him.

"Yes," she said, "but are we making full use of it?" Shapley paused—this sounded like criticism, after all—and he asked her what she meant. "At that moment an idea was born," she recalled later. "Why not use the Harvard plates to get all possible information about all the known variable stars?"[2]

Cecilia could identify the challenge in any scientific unknown. She was in gear again. Thus began an astronomical project that "was the most ambitious, complete, and thorough survey to have taken place in the first half of the twentieth century."[3]

Once again, the Harvard plates proved to be a gold mine. But devising a coherent classification scheme was daunting before the advent of computer crunching. Cecilia knew they had to prioritize the work. First, she and Sergei selected the variable stars that were bright enough to analyze. Then they divided them up by type. Sergei's doctoral thesis was on eclipsing binary stars (pairs of stars that brighten and dim as one moves in front of the other), so he took them. Cecilia took all the other types. With the help of a number of assistants, they eventually obtained almost two million observations of thousands of variable stars.[4]

Cecilia felt herself to be in unfamiliar territory. She had always gone it alone before, working at her own frantic pace, by herself at all hours in a darkened observatory. Now "I felt I was no longer a freelance, but a member of a team."[5] Yes, it was a team, but there was no question that Cecilia was the driving force. As Owen Gingerich, her student and a future Harvard astronomer, put it, she became "a relentless classifier of variable stars."[6]

But she didn't stop at just classification. She and Sergei probed the chemistry and physics of variables, their work providing clues as to how stars evolved and how galaxies formed. Gradually, Cecilia and Sergei turned variables into a valuable field of study.[7] In 1938 they published a coauthored book, *Variable Stars*, which presented their preliminary results. The vast project was completed fifteen years later.[8] "Variable stars were regarded as a sort of second-class problem," Cecilia wrote later. "Perhaps I have had a small hand in making them respectable."[9]

The day-and-night research at the Harvard Observatory could not mask what was happening abroad, however. The tension of a growing threat of war was darkening all of Europe, including the astronomical community. A conference on novae was scheduled for Paris in the summer of 1939. Cecilia was eager to attend; Eddington planned to be there. In typical fashion, she defied those who said it was unwise to go.

She was glad she went. She and Sergei stood in the crowd watching French troops as they marched gallantly past the Arc de Triomphe on Bastille Day. It was, for the moment at least, still very much gay Paris. And the final night of the conference provided a memory long kept. At dinner, Eddington—so formal, so proper, so old-school—turned to her and said, "Do you think your Husband would mind if I took you in to dinner?"

She smiled and took his arm—the man whose speech, transcribed word-for-word in a nineteen-year-old girl's hand, had inspired her to study the stars; the man who had seen "no insuperable objection" to her wishing to be an astronomer; the man she had felt so close to at Niagara Falls. It would be "the last time I ever saw the greatest man I have been privileged to know."[10] Eddington died of cancer in November 1944, a year before the war's end would allow transatlantic travel again.

It was only when Cecilia and Sergei traveled across the English Channel to London, the air thick with wartime tension, that another memory was recorded. "I shall never forget the panic scene in Whitehall, the crush of cars, people rushing to and fro between the Government offices." She and Sergei quickly booked passage on the SS *Normandie*; it would be the French liner's last voyage. "War was declared when we had been two days at sea," Cecilia recounted later, "and we made an eerie, blacked-out passage to New York."[11]

Back in the Harvard Observatory's seemingly safe surroundings, the beginning battles of a world war were just distant rumblings. In fact, a bit of progress had been made in recognizing the work of women at the observatory. Two of them had been promoted. In 1938, Annie Cannon and Cecilia were both appointed to positions with the title of astronomer. It wasn't "professor," but "it was a step forward for me," Cecilia wrote later, "for now I had a position, though still at a regrettable salary."[12]

Despite the regrettable salary, Cecilia was happy. Her marriage brought her companionship, common goals, mutual assistance. Other than when Norbert Wiener had courted her a full decade earlier, she had never experienced the kind of affection that Sergei showed her. At the observatory, she was all business; but privately with Sergei, and with close friends, she took delight in his attention. For his part, Sergei was genuinely proud of her accomplishments. And he needed her. When it came to writing letters and papers, his English was barely adequate.

"Routine work was a blessing," as Cecilia described it. "Variable stars and the responsibilities of a household filled our days."[13] A household that quickly began to grow. The shy and socially awkward college girl was now an internationally recognized woman astronomer in her mid-thirties with, hard to believe, a husband. Except for constant worry over money, they were a relatively carefree couple—but not for long. Only a year into the marriage, Cecilia and Sergei had a son, Edward, born in May 1935. A daughter, Katherine, followed less than two years later. A third child, Peter, was born in 1940.

With her thesis on the composition of stars, Cecilia had pushed against the boundaries of established science. Now, with her pregnancies, she pushed against other norms. She saw no need to curtail her work simply because she was carrying a child—a radical thought at the time. To raised eyebrows, she presented a paper to the American Astronomical Society when she was five months pregnant with her first child. She pushed the boundaries again with her second—a few months after Katherine was born, she left the baby with Sergei to lecture at the Yerkes Observatory at the University of Chicago for six weeks. But with her third child, she found the breaking point. Pregnant with Peter, she agreed to speak at Brown University. It would generate little more than a yawn today; but someone complained to Shapley, and he put his foot down. "The Observatory would be open to severe censure if you went to Providence in your present condition," he wrote to her in a March 1940 memo. She canceled her talk.[14]

And just as Cecilia had put stress on the men's club of science and on the social norms of the astronomical community at large, she put a strain on the observatory itself. When the children were very young, the combination of hers and Sergei's income provided enough funds to hire domestic help to look after the growing brood. But after December 1941, when the United States entered World War II, housekeepers and caregivers could find better paying work in the war effort. The solution: Cecilia took the kids to work. It was like take-your-son-or-daughter-to-the-observatory day, every day.

Shapley gritted his teeth and told the staff that they were expected to do the same, to tolerate the Gaposchkin children. Tolerate them after they found the dome of the 15-inch telescope, sat in the observer's chair, and rode it noisily up and down by madly turning the steering wheel. Tolerate them as they played hide-and-seek in the old dusty catacombs holding all those precious plates. Tolerate them as they wandered from Sergei's office—easy for them to enter, either through the door or, as often as not, through an open ground-floor window—to Cecilia's office, located in an isolated little nook, and reachable only by traipsing through a number of other offices.

Sergei's office and Cecilia's office were separated because Cecilia had inherited Henrietta Leavitt's desk a full decade before newcomer Sergei had arrived on the scene. It was an office layout that suited them both. "Mrs. G—that's what everyone called Cecilia Payne-Gaposchkin—seemed a formidable, rather remote presence," remembered Owen

Gingerich. "Sergei—that's how we all referred to her feisty husband— had his office [near] the 15-inch telescope, and the graduate students assumed that their two offices were widely separated to keep the stormy personalities from asserting themselves too conspicuously."[15]

It did not take long for Sergei to become a handful, in multiple ways. For one thing, he came to resent his status at the observatory. The price of passage out of Nazi Germany soon became apparent. Cecilia's pay was woefully small; Sergei's, however, was less than half of hers.

For another, although Sergei was energetic and hard-working, Cecilia's first impression that his analytical abilities were "good but not brilliant" was accurate. Sergei was an observatory employee, and so it fell to Shapley to deal directly with his less-than-stellar worker. When Sergei wanted to publish some of his findings in the prestigious *Astrophysical Journal*, Shapley had to level with him.

In an April 21, 1938, memo, he told Sergei that the reviewer "frankly does not like this particular paper—thinks it adds very little." He went on to write, ever so diplomatically, "I am inclined to suggest the withdrawal of the paper."[16] It would be hard to imagine Shapley ever having the need to write something like that to Cecilia.

Over and over, Sergei showed himself to be a loose cannon rolling around the deck of the Harvard Observatory. He was opinionated and rude to other staff members, but touchy about being criticized himself.[17] As director, Shapley maintained great patience, but it wasn't unlimited. After Sergei submitted a paper on variables that Shapley thought was "not happily organized," he pointed out to Sergei in another memo that "there are more variable star experts of one sort or another in the Harvard Observatory than in any other square mile or country on the face of the Earth. I leave it to you to derive whatever implications you think proper from that comment." He continued, an edge creeping in, "Sometime when the weather is right and the planets are in the right constellation you might want to discuss some of the points."[18]

Shapley was yet again in a tight spot. Sergei was the husband of one of his star employees, so he couldn't fire the man; but he also had to maintain his observatory's reputation and quality of work not only in the eyes of the astronomical community, but in the eyes of Harvard University. It was tightrope-walking that made for tangled prose. He opened one critical memo to Sergei with a warning: "If this note and the accompanying papers come to you on a hot day or when you [are] fragile in spirit, please put them aside and take the matter up again in a moment of calm."[19] He closed another memo by suggesting: "I think it would best not to bring the situation to [Cecilia's] attention at the present time when she is worried and wornout [sic]."[20]

After each of Shapley's barbs, Sergei would respond with a heartfelt and emotional apology. The replies were always handwritten. "I must tell you truth," he wrote to Shapley after a particularly critical memo, "that Cecilia has been almost in despair trying to prevent me from sending the paper to you before she had time to look at it."[21]

In another missive, Sergei apologized for apparently inserting himself into an issue between Shapley and Cecilia. He scribbled a hasty note to Shapley, admitting that "I was not right to interfere with Cecilia's private affairs . . . which brings for me only deplorable state of mind."[22]

No doubt Cecilia was "worried and wornout" because she was holding down two jobs—hers and his. Moreover, Sergei had placed her in her own tight spot. The man for whom she had fashioned a painstaking six-page case for bringing to America, the man for whom she had lobbied in London and Washington to receive a stateless person's passport, the man whom she portrayed as an eminent Russian astronomer was prone to writing her boss such inappropriate comments as, "I wish I could find a dark corner in which nobody could see me."[23]

Nor did it help Sergei's psyche for Cecilia to be in such constant demand. She received a steady stream of invitations to lecture or speak at colleges, observatories, seminars, and conferences around the country and the world. She rarely turned them down, routinely leaving the dishes and the kids to Sergei. By mutual agreement, she did the cooking and some cleanup; he paid the bills and did most of the housework.

Hers was not an easy role to play. She had only relatively recently moved to the United States from England, where there were only two paths for successful women: that of a single woman focused solely on her own career, or that of an educated wife who supported her husband's career.[24] Cecilia rejected both. Why could she not be simply a married scientist? Men were.

She could, but there was a cost. Winifred Holtby, a journalist and novelist in London in 1935, described what it was like for ambitious women of post-Edwardian England: "When a woman believes enough in her own mission to be ruthless . . . something happens. But most women dread before anything to 'cause an uproar' or inconvenience a family; and their work suffers."[25] In truth, it was not that different for ambitious women in America.

Cecilia was determined not to let her work suffer. If she had to experience the guilt of the selfish, so be it. "She was the more important scientifically, so she generally left [Sergei] as babysitter," recalled Cecilia's daughter, Katherine Haramundanis. "He in turn sometimes behaved flamboyantly, which, like the stoic she was, she tried to ignore."[26]

Sometimes flamboyant, sometimes more than that. Sergei was several inches shorter than Cecilia, with a bodybuilder's sculpted physique. "He was very proud of his legs, and wore super short shorts," remembers Robin Catchpole, an astronomer at the Cambridge Observatory, who once attended one of Sergei's presentations. "He hand drew colorful angelic figures in the borders of his slides. It was so odd. We were conservative back then, and he was very flirtatious with women around."[27]

Sergei made no secret of what he thought of various women staffers at the observatory. He would preen and flex, assuming they would be as impressed by his muscles as he clearly was. "In short, Sergei was a misfit in the astronomical community, to the extent that an astronomer might comment when he seemed to be acting normal."[28]

Partly acceding to the norms of the day, Cecilia grafted Sergei's last name on to hers. She was careful, though, to ensure that she be known as Cecilia *Payne*-Gaposchkin. She wanted it to be clear that the papers she wrote before and after her marriage were authored by the same person. But although she and Sergei now shared a name, Shapley and other astronomers often referred to her as *the* Gaposchkin.[29]

And *the* Gaposchkin was a whirlwind of work. By 1936, already a member of both the Royal and the American Astronomical Societies, Cecilia was elected to the American Philosophical Society, an honor followed a few years later by an honorary doctorate from Smith College. After *Stellar Atmospheres* and *Variable Stars*, she wrote another book, *The Stars of High Luminosity*. By 1942, she had written seventy-eight papers on stellar spectra, and another fifty-eight papers on stellar photometry. She continued to teach graduate-level astronomy and joined the Harvard Observatory Council. Working at her cluttered desk, she then focused her attention on the Magellanic Clouds, two southern sky galaxies orbiting the Milky Way. She and Sergei made more than two million individual measurements of variable stars within the galaxies from photographic plates at the observatory.[30]

All that was not enough, however, to change her status at Harvard. Her "regrettable salary" continued to hover at $2,700 a year, and her name was still missing from the Harvard course catalog. In looking back at those years, Dorrit Hoffleit, a research astronomer at Yale (who had been at Harvard, except for the war years, from 1930 to 1956), described Cecilia as "the most brilliant and at the same time the person most discriminated-against at Harvard College Observatory."[31]

Much of that discrimination can be traced back to the Dear Director. Unbeknownst to Cecilia, Shapley fended off opportunities that became available to her. In March 1941, a classics professor at Bryn Mawr wrote to Shapley asking what he thought about the possibility of making Cecilia president of the college. Shapley was not inclined to help an effort to take away a prized and inexpensive resource. In his reply, he gave three reasons for not being able to recommend Cecilia: (1) she was too good an astronomer to be diverted into administrative duties; (2) her "genius temperament" might not allow her to get along with more conventional professors; (3) the Gaposchkin family was unconventional—to wit, Sergei's personality could cause problems.[32]

Sergei's reputation in the tight-knit astronomical world was clearly a drag on Cecilia's chances for advancement. Other astronomers were quick to cite him as a reason for side-stepping Cecilia. When her name surfaced as a potential secretary of the American Astronomical Society, Russell nixed it, noting that Cecilia was already pushed to the limit with her "professional and domestic obligations."[33]

When Otto Struve, director of the Yerkes Observatory, was asked to suggest candidates for a professorship at Harvard, he listed Cecilia, but qualified it by writing that her "domestic situation might be a drawback."[34] But when Russell was asked to recommend Cecilia for a Radcliffe professorship, he made sure to note that she was the more capable half of the couple: "She is his equal in industry . . . and his superior in knowledge and judgment."[35] The appointment went to someone else.

Cecilia was aware of the ripples Sergei was stirring up in the observatory and beyond; they weren't hard to see. But she must have realized that the reins on her career were related less to Sergei than to gender. She would from time to time lash out, but it was in typical Cecilia fashion, combining perception with edgy humor. In a 1937 talk before the American Association of University Women, she noted that "the woman scholar today is expected to live the life of a recluse, as was required of men scholars a hundred years ago."[36]

Not to say that there weren't times that relieved the professional and personal stress. The Harvard Observatory was at the crossroads of astronomical thought, and scientists of all kinds were always passing through. But one visitor was welcome precisely because he was not part of the astronomical community. Cecilia's brother, Humfry, sailed to America in 1935. He had taken time off from his archaeological excavation work in Greece because he wanted to see his new nephew, who carried the name of his father: Cecilia and Sergei's six-week-old son, Edward.

Sergei, Cecilia, Humfry, 1935

Cecilia was thrilled to see him; true family contact was rare. She had not seen Humfry for many years, and she must have marveled at how the man could find suits to fit; he was six foot eight now, a gentle giant. He towered over five-foot-ten Cecilia; and Cecilia in turn towered over Sergei. A photograph made during the visit shows the three of them, standing together on a rocky cliff, a study in contrasting heights.

Brother and sister shared much during Humfry's stay in the Payne-Gaposchkin cramped apartment: laughs, laments, memories. And cigarettes! Humfry had given Cecilia her first cigarette during their London days, and almost every picture during Humfry's brief visit shows a cigarette in the hand of one or the other.

Cecilia would come to treasure those photos. It was the last time she would see her brother; Humfry would be dead within a year. He contracted a staph infection when he returned to Athens and died on May 9,

Humfry lights Cecilia's cigarette, 1935

1936. He was thirty-four years old; his death came one day before Cecilia's thirty-sixth birthday.

To put some separation between herself and the frustration and flamboyance swirling around her, Cecilia convinced Sergei that they needed to move their family of five to a larger house. They found a two-story white clapboard house on Shade Street in Lexington, complete with an English-style country garden with flowers showing through a picket fence. The mind that relentlessly classified stars was as restless as ever; but here, miles from Cambridge, Cecilia could indulge herself in interests far from the outer reaches of the universe. She sewed with the precision of an expert seamstress. She tried her hand at soap making. She took home colorful printouts of spectra, laid out in meticulous mathematical grid patterns, and then created needlepoint replicas on a black background with balls of yarn in shades of red, orange, blue, green.

Evenings in Lexington were centered on the kitchen. She enjoyed trying different cooking styles, from French to British to Asian. "Such was her enthusiasm for cooking," recalled her daughter Katherine, "that after every dinner she prepared the kitchen was a shambles."[37]

Money was always tight, which meant she and Sergei needed to be inventive, although Cecilia might have carried frugality to an extreme. She made quince jelly from her garden's tree, straining the pulp with a pillowcase hung between two kitchen chairs.[38] And then there was the parachute. She spied one in an army surplus store and immediately bought it, turning its rare and expensive yards of gossamer white nylon into underwear for the kids.[39]

Sometimes—not often, but sometimes—her unceasing need to feed her mind would be on display in the observatory. Those who got to know her well enough might get a glimpse of her wide range of nonstellar interests. Her graduate student Owen Gingerich was one. He marveled at "her views on Italian art or paleolithic axes or mosaic woodworking or the earliest printed edition of *Reynard the Fox*, all topics that deeply interested her."[40]

A few years after the Lexington house purchase, she and Sergei added a small hardscrabble farm to their real estate holdings. It was located in the rocky soil of northern Massachusetts, and it was primitive. The family lived in a one-room cottage that Sergei built in the pine woods. Food was cooked on a small wood stove (Cecilia called it her "pet"). Water was hauled from half a mile away. There was no electricity or indoor plumbing.[41]

Frugality ruled here, too. Cecilia was maniacal in her canning—poring over government bulletins, inspecting each bottle, boiling everything in sight. Albert, the sheep tethered in the back yard, was utilized as a mower; in time, he ended up in jars. The two piglets were turned into bacon and spare ribs. Sergei slaughtered the chickens by wringing their necks with his bare hands.

Life was physically hard. The kids had chores, too; they collected eggs early every morning. But they were happy days. Although no one would ever describe Cecilia as being totally satisfied, she was more or less content with her life. She wore her hair short; she rarely wore makeup. Once the children were in bed—always with "something to read"—she would unwind with a deck of cards and play solitaire, Sergei kibitzing at her side. "She never followed the foibles of fashion," says her daughter Katherine, "nor engaged in time-consuming attempts to retain youth."[42]

Life outside the home and the farm, however, was not so peaceful. The world was at war, and with Americans on the front lines and London under daily attack, both Cecilia and Sergei felt the need to get involved. At first it was through donations. In their spare time, they would drop off hundreds of eggs and several turkeys to charity. They bought a horse and buggy to save fuel.

But as the fighting in Europe worsened, they edged into politics. They founded the Forum for International Politics. Shapley provided a lecture hall at the observatory, and Cecilia and Sergei began to hold weekly meetings. They had no trouble getting speakers; the university and the local community were teeming with people who wanted their say on England, the United States, France, Germany, Russia, India, Palestine. The gatherings were supposed to promote "enlightenment," but they quickly dissolved into raucous debates. Cecilia was the chair, always struggling to be dispassionate when she really wanted to be partisan, occasionally fearing physical violence on the platform. And, such as were the times, the forum earned her "the reputation of being a dangerous radical."[43] Years later, Sergei was called before a traveling arm of Senator Joseph McCarthy's House Un-American Activities Committee when it visited Boston.[44]

Music would provide balance to politics. Some was a bit offbeat: late at night, Cecilia would hear the night assistant, Frank Bowie, as he belted out Guy Lombardo's rendition of *Charmaine*, his voice bouncing off the walls of the stacks as he developed the glass plates. Some was a bit more professional: Cecilia founded the amateur but enthusiastic Observatory Philharmonic Orchestra. Once again Shapley was called on to provide space. Cecilia organized regular rehearsals, headed the violin section, and—Gustav Holst would have beamed—conducted the first performance. Frances Wright brought in a baby grand, and, with Cecilia holding the baton, banged out a credible rendition of a piano concerto written by astronomer-composer Sir William Herschel.

Cecilia was even pressed into opera duty. It did not go well. Entitled the "Observatory Pinafore," it was a takeoff on Gilbert and Sullivan, poking fun at Pickering's early photometric work. The performance took place during a meeting of the American Astronomical Society. Shapley demanded that the entire observatory staff participate. "I was promoted to the chief woman's part," Cecilia recalled. "Alas, it is a soprano part, and I was a contralto. No matter, it had to be done, with the result that I have never been able to sing since."[45]

At the time, though, there were more concerns to deal with than just musical performances and politics. Chief among them, the children. With both parents working six days a week—and with many of those days stretching into night—there wasn't much time left for attention. But there was more than just a self-imposed intense work schedule competing for family time. Cecilia was by nature a stoic person, at times seeming to lack any emotion. "I never saw her cry," recalled her daughter Katherine. In fact, "it was a rarity to hear her laugh out loud."[46]

That void of attention produced predictable results. Edward and Katherine, looking for some notice, could be a bit rambunctious in the eyes of the observatory staff. But it was the youngest of the three Gaposchkin siblings who reacted most strongly. What little attentiveness Sergei and Cecilia had the capacity for had to be spread among three children, and poor Peter's response was—no response. He simply did not talk. His silence created more and more anxiety for both Cecilia and Sergei.

Doctors were not sure if he was deaf, intellectually disabled, or just plain fearful.[47]

In 1943, when Peter turned three years old, still speechless, Cecilia, herself close to a breakdown, hired the wife of astronomer Dean McLaughlin to care for Peter during the day. The woman's "casual, cheerful kindness" clicked somehow with the little boy.[48] Peter startled the entire family one morning by running downstairs singing "From the Halls of Montezuma" at the top of his lungs. It was a great relief, "but scars were carried for many years from those early anxieties," wrote Katherine.[49]

Amid all the family tumult, however, there was one bright spot— another crack in the all-male Harvard faculty. Cecilia's honors and papers—248 at this point—had piled up so much that, finally, a student in 1945 perusing the university course catalog would discover that the introductory course in astronomy was taught by "Cecilia Payne-Gaposchkin, Astronomer." A small detail, but a big deal—another sign that in time the dam would break.

More change was on the way. In the course of three decades, Shapley had accomplished what Cecilia and Adelaide Ames had forecast: the Dear Little Observatory had indeed become a Great Institution. But by the early 1950s, the university was making noises that he should think about retirement. Years earlier, he had said to Cecilia, "We must all ask ourselves, have we picked our successors?"

At the time, Cecilia had replied, casually and unrealistically, "I don't mind your leaving, if I can succeed you." Now, with the moment at hand, she knew it would not happen. She rationalized it by thinking to herself that she would have clung to the past, when "the institution that he had built must respond to the changing times and the changing face of science."[50]

Shapley was focusing more and more of his energy on national and international issues, raising money to support liberal candidates for political office, organizing peace rallies. Conservatives in Congress "took a dim view of [his] activities, and he was made the subject of nasty and difficult investigations by several congressional committees."[51] With his effectiveness as leader of the observatory on the wane, what was obvious to everyone soon came to pass: in 1954 Donald Menzel became the seventh director of the Harvard College Observatory. (He had been appointed acting director two years earlier amid considerable internal turmoil, but he had strong support from Cecilia.)

One of Menzel's first tasks was to bring some discipline to the halls of the observatory. Cecilia's youngest child, Peter, was now a teenager. As he grew older, he had progressed from not talking at all to talking too much. In 1958, the Harvard Observatory Council met and officially voted to warn Peter to stop disturbing the staff in the observatory library.[52] Cecilia was out of town for the meeting. When she returned and saw that the vote was recorded in the official meeting's minutes, she was furious.

In a letter to the council, she wrote that the vote in her absence "has hurt me deeply. Do you feel that it is fit treatment for one of your own number, who is devoting her whole time and energy to the good of the Observatory?"[53] The irony that Peter's troubles might stem in part from the very devotion she described was not noted.

Cecilia wrote in the note that Peter was "ready, and I think is entitled, to hear the criticisms personally." And so Menzel did just that. In two extraordinarily formal two-page typewritten "Dear Peter" letters, Menzel requested that the boy stop demanding time from busy staffers and stop interrupting their important work. "You are now a very large person," Menzel wrote. "You have a tendency to lean over a person and then put your face right in his face. This is both awkward and somewhat distasteful. Can you not learn to stand erect. Very erect."[54]

Apparently, Peter had found ways to converse even when there wasn't another person around. To make sure Peter got it, Menzel spelled it out: "You will endeavor to break that bad and annoying habit of talking and laughing to yourself which other people find juvenile and distasteful." Menzel then pivoted from critic to businessman, closing his letter to the teenager with: "If you agree to the above [restrictions], will you please sign the enclosed carbon, as indicated, and return it to me in the enclosed envelope."[55]

As Menzel's tone makes clear, there was a breaking point in the observatory's collective patience. From Sergei's flamboyance to Peter's inability to restrain himself, there was a self-discipline problem in the Gaposchkin family. And much of it could be laid at Cecilia's feet.

This eminent astronomer who used earthly science to probe and understand distant phenomena could be, herself, so distant. She would always say, "'Let's talk about it later,'" remembers Katherine, "a 'later' that never came."[56]

The little girl who stared in wonder at a meteor; the eight-year-old who felt a life's calling in recognizing the bee orchis; the teenager who saw in her school's laboratory "the warp and woof of the world"; the determined college fresher who told a grouchy old biologist how wonderful it would be to do research; the graduate student who walked on air as she homed in on discovery—throughout her life, Cecilia was completely consistent. In a contest between science and anything else—politics, money, social expectations, family life, a child's homework—Cecilia would always choose science.

That is not to say that she always ran away from family issues. Her look could be stern, and often that was enough. There was no mistaking when she was displeased. But more often, she would call on her wit. When the household noise rose to the level of breaking her concentration, for example, she would smack the kitchen table with her open palm and exclaim, "I want silence! And very little of that!"[57]

Eventually, the children flourished. Edward (known as Mike) earned a doctorate in geophysics at Harvard and worked at the Smithsonian Astrophysical Laboratory; Katherine graduated from Swarthmore, also worked at the Smithsonian, and became a technical writer; Peter graduated from MIT and received a PhD in astrophysics from the University of California, Berkeley.

Menzel's next task after imposing some order was to examine the observatory's books; when he did, he was shocked to find out how little

Cecilia with her daughter Katherine and son Edward

Cecilia was paid. The man from whom she had been so estranged, the enemy in a Shapley-arranged truce for so many years, raised and then doubled her salary.[58] And with Shapley gone and Abbot Lawrence Lowell no longer president of the university (he had retired in 1933), Menzel took yet another long overdue step.

Cecilia's daughter Katherine was away at school. There was no phone call or letter from her mother; the first word Katherine received was when she read about it. "In those days," she wrote, "I was not attuned to the difficulties faced by a woman in the professions, particularly at Harvard, a bastion of maleness, but it seemed nonetheless a very great achievement."[59]

It was more than a great achievement; it was historic. The *New York Times*, June 21, 1956, reported the news: "Harvard University announced today the appointment of Dr. Cecilia Payne-Gaposchkin as Professor of Astronomy. She is the first woman to attain full professorship at Harvard through regular faculty promotion."

Harvard University, America's oldest, was founded in 1636 as an all-male institution of learning. There had always been women on campus, but "as workers and donors. As invisible helpmeets to fathers, husbands, and sons—not only as life's mainstays but also as intellectual collaborators. . . . In the late 19th century Harvard saw itself as the nursery for the nation of leadership and scholarship. Although Harvard men were generally sympathetic to the idea of educated women, who as Mothers of the Republic and teachers bore great responsibility for the young, they did not want women to study at the sacred grove reserved for the future ruling elite and intelligentsia."[60] The faculty had been as male-dominated as the names on the residential houses along the Charles River.

Even Cecilia, normally so humble, recognized the significance of this event and decided to mark it with a celebration. She sent out a handwritten invitation to every woman student of astronomy to join the celebration in the observatory's library. When she spoke, it was with classic self-deprecation. "I find myself," she told the assembled crowd, "cast in the unlikely role of a thin wedge." Cecilia was almost six feet tall, "broad shouldered in an era that valued petiteness."[61] It brought down the house.

So many Harvard firsts. The first PhD in astronomy, the first winner of the Cannon Award, the first woman promoted to professor at Harvard—and, a few months later—the first woman at Harvard to chair a department. Once a lowly woman graduate student, now Cecilia was professor and chair of the Astronomy Department. "I have reached a height," she wrote later, "that I should never, in my wildest dreams, have predicted fifty years ago."[62]

As the languid fifties became the tumultuous sixties, and the sixties the seventies, both Harvard and its first woman professor promoted from within the university and the first woman department chair experienced many changes. Some of the changes were welcome. Her increased salary, for example, certainly eased the household's money strain.

Other changes, though, were not so welcome. The demands on her time skyrocketed. The ever-increasing teaching burden and the managerial responsibility for a large and growing graduate school left no time for research. She was philosophical about it, taking pride in her department as it churned out fellow scientists: "As a mother sees her life renewed in her children, I saw my scientific efforts perpetuated in my students."[63]

The department now was her focus, the newest challenge to be attacked. It took its toll. "The responsibilities of the job put her under a great strain," observed her daughter Katherine, "which she controlled with cigarettes, coffee and the occasional pill."[64]

Astronomy was changing too, rapidly. High-energy astrophysics labs were probing ever deeper into particle research, and satellites were transporting telescopes from the observatory dome into deep space itself. At the Harvard Observatory, there was no better symbol of that change than the storied director's residence. In the early 1960s, the sprawling labyrinthine brick house fell to a wrecking ball. A sleek modern office building now occupies "that once romantic spot."[65]

But in the swirl of construction and invention, there were also things that did not change. Cecilia's desk at the observatory continued to be a wreck, covered with stacks of papers. Whenever a staffer would pass by, stop and stare, then make some comment about the piles, she would always respond, somewhat crossly, "My office may be disorderly, but my mind is not!"[66]

And on a corner of that desk, in its own reserved place, her ashtray continued to overflow.

20

Cecilia was fading.

Ever the adventure-seeking traveler, in the summer of 1979 she and Sergei had circumnavigated the globe, marking her seventy-ninth birthday: Tahiti, Australia, the Philippines, India, Turkey. She had marveled at the cathedral of Hagia Sophia in Istanbul and the Temple of Diana in Ephesus.

But it was a tiring trip, and she came back exhausted. She found it hard to breathe. She entered a Cambridge hospital for a biopsy. Fifty years of chain-smoking had finally exacted its toll. She had advanced lung cancer. She endured the requisite radiation treatments, but she grew weaker and weaker. Although her body may have been slipping, however, her will was not. She never complained.

At first, confined to home but restless as always, she could move from the second-floor bedroom to the first-floor living room. Soon though, even that was too much. As 1979 came to a close, she was moved to the hospital full-time. Her daughter Katherine was always nearby, softly playing her favorite music, Mozart's *The Magic Flute* and Handel's *Messiah*, through a small tinny speaker placed on her pillow.[1]

As she sat with her mother, Katherine recalled a decades-earlier trip she and Cecilia had taken together.[2] It was just the two of them in 1949, a mother exposing her daughter to learning far from the confines of a classroom. They had spent a week in London. Mother and daughter then crossed the Channel to Paris, where they stayed in a fine old hotel, and where wide-eyed Katherine attended the Paris Observatory's annual elegant champagne reception at Versailles.

It was when traveling that Cecilia would give a rare nod to fashion. She took along far too many clothes and was forced to lug heavy suitcases through railway concourses as she and Katherine made their way to Italy. At the open-air Loggia dei Lanzi in Florence, Cecilia, always the professor whether titled or not, pointed out that Benvenuto Cellini had taken just as much care sculpting the back of Medusa's head as he had her face, even though a viewer rarely saw the back. She said it was the mark of a true craftsman.

At Pisa, inside the Leaning Tower's cathedral, Cecilia explained that the bronze lamp hanging from a long chain had given Galileo the idea for a pendulum. She then took her daughter to see the museum housing the telescope Galileo had used to observe the moons of Jupiter. "It was amazing," Katherine later recalled, "that such a tiny instrument, smaller than a modern spyglass, could have so changed mankind's world view."[3]

As Cecilia lay in the hospital bed, tributes were written and friends paid respects; but it was family that remained close and constant: Katherine, her brothers Edward and Peter, and Sergei. Whatever troubles Sergei's professional and personal behavior had caused were pushed aside. Cecilia had signaled as much. She had dedicated her last book, *Stars and Clusters*, which she had finished only a month before, to her husband, "that bright particular star."[4]

One tribute, however, did not make it in time. Vera Rubin, a pioneering astronomer at the Carnegie Institution and a tireless advocate for women

Cecilia and her daughter Katherine touring France and Italy, 1949

scientists, always maintained that it was "an enormous injustice" that Cecilia had never been elected to the National Academy of Sciences. Although Cecilia had been first in so many ways, the academy consistently had excluded her, as it had other distinguished women scientists. When Vera asked if the Harvard Astronomy Department would nominate Cecilia for membership, she was initially rebuffed. Finally, Leo Goldberg, director of the Kitt Peak National Observatory, agreed to nominate her.[5] Whether she would have been voted in or not would never be known. Cecilia died on the morning of December 7, 1979, just before the election was to be held.

In her obituary in the *Quarterly Journal of the Royal Astronomical Society*, Cecilia was described as "a pioneering astrophysicist and probably the most eminent woman astronomer of all time."[6] Cecilia surely would have bristled at that. Not the "probably" part—she was genuinely humble and would not have quibbled with any qualifier of the world "eminent."

It would be the "woman" part she would have had an issue with. All her life she would ask others, including herself—what does gender have to do with scientific scholarship? She did not consider herself a woman astronomer; she was an astronomer.

In that sense, astronomy was the perfect match for her. Stars, and the elements of which they are made, did not care about gender. Stars did not discriminate; they were open to being friends with anyone. Any woman, any man, who found the key could unlock their secrets.

Although science would preoccupy her from the day she first observed "a shining meteorite," her mind had great capacity for things nonscientific. She fused the literary with the scientific to make the scientific so much more literate. Her title for a 1957 speech to women science scholars was "No Wine So Wonderful as Thirst." She took it from Edna St. Vincent Millay's poem about how need is preferable to satisfaction. She closed the speech by noting that it was a poet, not a scientist, who captured the essence of astronomy. She quoted from Laurence Binyon's poem "Thinking of Shores That I Shall Never See":

> Knowing how unsure is all our knowledge, doled
> To sloven memory and to cheated sense,
> And to what majesty of stars I hold
> My little candle of Experience.

In that same speech, Cecilia made an admission to the assembled audience: "I confess that I personally owe more to Odysseus and Nausicaa than to Einstein and Kepler."[7]

It was that very ability to absorb and understand so easily and quickly that put her into conflict with established astronomical thought. The huge amount of data embedded in the photographic plates at the Harvard Observatory, collected over decades by a dedicated team of women computers, had not been fully utilized because no one had yet figured out how to interpret it. Cecilia's knowledge of the new physics, along with her intelligence and doggedness, was required. But she also faced headwinds:

the "uniformity of nature" consensus among astronomers of the time. Most astronomers of the previous generation were not formally trained in this "new physics" and were quite naturally threatened by it.

Cecilia's relationship with the men of science was a complicated one. For sure, men were obstacles, but they were also enablers. At Cambridge, Searle had demanded that she and other women take off their corsets, but he also ingrained in her the importance of precise measurement. Bateson reduced her to tears, but her appreciation of his systematic approach to research helped her make sense of hundreds of thousands of photographic plates. Rutherford derided her in class, but he taught her atomic theory and infused in her his get-it-done discipline. Eddington dismissed her careful and thorough thesis results, but with his eloquent lectures, he inspired her to become an astronomer. Shapley diverted her from her "beloved spectra," undeniably to the detriment of science, but he also gave her the opportunity for discovery and provided a job for her husband. Menzel had competed with her, but he also paid her a fair salary and made her the first woman promoted to tenured professor at Harvard. And Russell had characterized her remarkable breakthrough about the abundance of hydrogen as "seriously wrong," only to later discover its truth himself; but he remained a steadfast friend, confidant, and, most important, fellow scientist.

Cecilia had her own way of dealing with learned men who stood in the way. She absorbed their wisdom, and then simply went around them. She did not look to them for approval in order to continue; she looked higher. Whether it was religion or science—God or Universe—she raised her sights above the mere mortal and focused on the source. "I have always been in direct touch with the fountain-head," as she described it. "No other mortal has made my intellectual decisions for me. I may have been underpaid, I may have occupied subordinate positions for many years, but my source of inspiration has always been direct."[8]

Going direct served her well. It was a way of thinking and acting that enabled her to step over gender-generated obstacles her entire life. In 1970, long after she had proven herself with honors and awards and recognition and titles, a letter written by Harvard's dean of freshmen, F. Skiddy von Stade Jr., was reprinted in the *Harvard Crimson*: "Quite simply, I do not see highly educated women making strides in contributing to our society in the foreseeable future. They are not, in my opinion, going to stop getting married and / or having children."[9]

Cecilia maintained a consistent response to these kinds of comments—always adding a touch of humor, always proclaiming faith in her long-held principle of patience: "The truth will prevail in the end. Nonsense will fall of its own weight, by a sort of intellectual law of gravitation."[10]

It helped being in America. If she had remained in England, it is doubtful that she would have had enough access to the equipment, data, and other scientists needed for a successful scientific career. She had entered the United States on a visa but stayed a lifetime. Harvard in particular, and the country as a whole, allowed her to apply her own version of evolution to the task of discovery. "It has been a case of survival, not of the fittest, but of the most doggedly persistent," she wrote later. "I was not consciously aiming at the point I finally reached. I simply went on plodding, rewarded by the beauty of the scenery, towards an unexpected goal."[11]

Not that she wasn't human. She carried around her own supply of foibles. She could be jealous. As a young girl, she coveted her brother Humfry's apparently easy glide through life, from carriage rides to school admissions. It irritated her that her colleagues at the Harvard Observatory sometimes had more access than she did to the director, Shapley. Despite her personal credo to put scientific advancement before personal feeling, she still fumed when Harry Plaskett got the offer from Oxford.

She could be temperamental. A young woman who worked for Cecilia reported that "once, in a fit of temper, Miss Payne had picked up a pile of glass photographic plates and smashed them to the floor!"[12] It's hard to believe that she actually threw one of her treasured plates, but no doubt she threw something.

But foibles are forgivable when one considers what an extraordinary person she was. She was an explorer every bit as fearless as those history has celebrated. She enjoyed probing the heavens in her mind the same way she enjoyed exploring the world in person. Jesse Greenstein, her student at Harvard in 1927 and later a professor at the California Institute of Technology, described it as well as anyone. He wrote that her body of work—the papers, the books, the speeches, the lectures—"showed the bravery and adventure of a mind exploring the unknown with the available scientific apparatus and a complete belief in the power of human reason and logic."[13]

It would take many years for the astronomical community to fully acknowledge what Cecilia had accomplished. In 1962, when the astronomer Otto Struve pronounced Cecilia's *Stellar Atmospheres* "the most brilliant thesis ever written in astronomy," it had been thirty-seven years since it had been published. He was late, but he was correct. In a two-year stretch of intense research, she had made a fundamental scientific discovery about the universe. But like a gifted rookie athlete who goes on to have a long career, Cecilia followed her discovery with 284 publications, eight of which were coauthored with Donald Menzel after he became chair of the department.

There is an honor that the American Astronomical Society awards annually "on the basis of a lifetime of eminence in astronomical research." Henry Norris Russell was the first one to receive the prize, in 1946. The prize had never gone to a woman until Cecilia won it in 1976, three years before her death—another first. No doubt she appreciated the irony; the award was then, and still is, known as the Henry Norris Russell Prize.

In the opening of her acceptance speech, in ringing words as fundamental as her discovery fifty years earlier, she contrasted the one-night occasion of winning an honor with the decades of reward a scientific career had provided her: "The reward of the young scientist is the emotional thrill of being the first person in the history of the world to see something or understand something. Nothing can compare with that experience. . . . The reward of the old scientist is the sense of having seen a vague sketch grow into a masterly landscape."[14]

Today, Cecilia's "masterly landscape" seems almost quaint. The "quickening influence of the Universe," as she described it, is immeasurably quicker. In its day, the 15-inch Great Refractor, deep inside a dome at the Harvard College Observatory, was the king of telescopes. Now, telescopes many times larger orbit the earth. Photographic glass plates have given way to far more efficient computer-generated images.

And measurement today has attained a level of precision well beyond what was practiced at the Cavendish Lab. A billion years ago, the collision and merging of two black holes set off a gravitational wave. When that wave arrived on earth—on September 4, 2015—it created an interference in laser detectors located in Louisiana and Washington. The interference was just a flicker, but it was enough for scientists to be able to convert it into a recognizable signal—a split-second chirp. One can imagine Cecilia marveling at so precise a measurement, a deviation a fraction of the width of one of Rutherford's protons, detectable only with enormous computing power.

And the belief that the universe is mostly hydrogen? We know now that most of the atoms in the universe are indeed hydrogen atoms. But there is other stuff out there. So-called dark matter has five times more mass than that of stars. The best guess today is that it is composed of some kind of elusive subatomic particle or particles.

One thing has always remained constant, however: the simple thrill of discovery. Sometimes the thrill is spread in headlines worldwide. More often, it's contained in the office, the lab, or the observatory. Although Cecilia and her work over the years came to be well known in the astronomical world, recognition did not extend beyond that tight-knit community. Shy as a young girl, and later intensely focused on her work, she was not a self-promoter, was not one to "breeze up." It makes it a challenge to satisfy the need to understand this woman who so needed to understand. To truly know Cecilia, the best words to turn to are Cecilia's.

When young people, especially young women, would ask for advice, she had a ready answer. The words are directed specifically to those looking to turn to science; but between the lines, they allow us to know

"Harvard University announced today the appointment of Dr. Cecilia Payne-Gaposchkin as Professor of Astronomy. She is the first woman to attain full professorship at Harvard through regular faculty promotion." *New York Times,* June 21, 1956.

Cecilia herself. They are a scientist's words to other scientists, but they speak to anyone who, confronting the unknown, must call on that need to understand that the human spirit is so capable of:

> Do not undertake a scientific career in quest of fame or money. There are easier and better ways to reach them. Undertake it only if nothing else will satisfy you; for nothing else is probably what you will receive. Your reward will be the widening of the horizon as you climb. And if you achieve that reward you will ask no other.[15]

Notes

Author's Note

1. "Portrait of a Pioneer," *Harvard Magazine*, March–April, 2002.
2. Quoted in Herbert F. Vetter, "Cecelia Payne-Gaposchkin: Astronomer and Pioneer," *UU World*, January 1, 2003, https://www.uuworld.org/articles/cecelia -payne-gaposchkin.
3. Richard Fortey, *Life: A Natural History of the First Four Billion Years of Life on Earth* (New York: Knopf, 1997), 13–14.

Prologue

1. Dorrit Hoffleit, "Reminiscences of Cecelia Payne-Gaposchkin (1900–1979)," in *The Starry Universe: The Cecilia Payne-Gaposchkin Centenary*, ed. A. G. Davis Philip and Rebecca A. Koopmann (Schenectady, NY: L. Davis Press, 2000), 91.
2. Cecilia Payne-Gaposchkin, "The Dyer's Hand: An Autobiography," in *Cecilia Payne-Gaposchkin: An Autobiography and Other Recollections*, ed. Katherine Haramundanis (Cambridge: Cambridge University Press, 1984), 163.
3. Author's correspondence with Professor Virginia Trimble, Physics Department, University of California, Irvine, October 2018.
4. David H. DeVorkin, "Quantum Physics and the Stars (IV): Meghnad Saha's Fate," *Journal for the History of Astronomy* 25 (1994): 158.
5. Payne-Gaposchkin, "The Dyer's Hand," 136.
6. David H. DeVorkin, "Extraordinary Claims Require Extraordinary Evidence: C. H. Payne, H. N. Russell and Standards of Evidence in Early Quantitative Stellar Spectroscopy," *Journal of Astronomical History and Heritage* 13, no. 2 (2010): 139.
7. Otto Struve and Velta Zeberg, *Astronomy of the 20th Century* (New York: Macmillan, 1962), 220.

8. Richard Williams, "January 1, 1925: Cecilia Payne-Gaposchkin and the Day the Universe Changed," *American Physical Society News* 24, no. 1 (January 2015).

9. Owen Gingerich, Leo Goldberg, Fred I. Whipple, and Charles A. Whitney, "Cecilia Helena Payne-Gaposchkin: Memorial Minute," *Harvard University Gazette* (May 8, 1981): 6.

Chapter 1

1. Cecilia Payne-Gaposchkin, "The Dyer's Hand: An Autobiography," in *Cecilia Payne-Gaposchkin: An Autobiography and Other Recollections*, ed. Katherine Haramundanis (Cambridge: Cambridge University Press, 1984), 86.

2. Norbert Wiener to Constance Wiener, July 21, 1925, MC-0022, box 2, folder: Correspondence, January–July 1925, Norbert Wiener Papers, Department of Distinctive Collections, MIT, Cambridge, MA.

3. August Wenzinger, "The Revival of the Viola da Gamba: A History," in *A Viola da Gamba Miscellany: Proceedings of the International Viola da Gamba Symposium*, ed. Johannes Boer and Guido van Oorschot (Utrecht: Foundation for Historical Performance Practice, 1991), 134.

4. Payne-Gaposchkin, "The Dyer's Hand," 96.

5. Payne-Gaposchkin, "The Dyer's Hand," 79, 82.

6. Payne-Gaposchkin, "The Dyer's Hand," 86–87.

7. Carol Dyhouse, *Girls Growing Up in Late Victorian and Edwardian England* (London: Routledge, 1981), 23.

8. Payne-Gaposchkin, "The Dyer's Hand," 85–86.

9. Katharine Chorley, *Manchester Made Them* (London: Faber and Faber, 1950), 150.

10. Payne-Gaposchkin, "The Dyer's Hand," 87–88.

11. Henry Tyrrell, *The History of the War with Russia*, vol. 2 (London: London Printing and Publishing Company, c. 1857), 356.

12. Payne-Gaposchkin, "The Dyer's Hand," 86.

13. *The Law Times*, December 31, 1904, 213.

Chapter 2

1. Cecilia Payne-Gaposchkin, "The Dyer's Hand: An Autobiography," in *Cecilia Payne-Gaposchkin: An Autobiography and Other Recollections*, ed. Katherine Haramundanis (Cambridge: Cambridge University Press, 1984), 84–85.

2. Payne-Gaposchkin, "The Dyer's Hand," 89–91.

3. Payne-Gaposchkin, "The Dyer's Hand," 91.

4. Payne-Gaposchkin, "The Dyer's Hand," 92.

5. Clement John Wilkinson, *James John Garth Wilkinson: A Memoir of His Life, with a Selection of His Letters* (London: Kegan Paul, Trench, Trübner, 1911), 95.

6. Payne-Gaposchkin, "The Dyer's Hand," 92.

7. Payne-Gaposchkin, "The Dyer's Hand," 94.

8. Payne-Gaposchkin, "The Dyer's Hand," 93.

9. Payne-Gaposchkin, "The Dyer's Hand," 90.

10. Payne-Gaposchkin, "The Dyer's Hand," 92.

Chapter 3

1. Cecilia Payne-Gaposchkin, "The Dyer's Hand: An Autobiography," in *Cecilia Payne-Gaposchkin: An Autobiography and Other Recollections*, ed. Katherine Haramundanis (Cambridge: Cambridge University Press, 1984), 94.

2. Carol Dyhouse, *Girls Growing Up in Late Victorian and Edwardian England* (London: Routledge, 1981), 44.

3. Payne-Gaposchkin, "The Dyer's Hand," 96.

4. Payne-Gaposchkin, "The Dyer's Hand," 97.

5. Payne-Gaposchkin, "The Dyer's Hand," 104.

6. Payne-Gaposchkin, "The Dyer's Hand," 104.

7. Payne-Gaposchkin, "The Dyer's Hand," 98.

8. Payne-Gaposchkin, "The Dyer's Hand," 98.

9. Payne-Gaposchkin, "The Dyer's Hand," 97, 110.

10. Payne-Gaposchkin, "The Dyer's Hand," 97.

11. Payne-Gaposchkin, "The Dyer's Hand," 99.

12. Payne-Gaposchkin, "The Dyer's Hand," 101.

13. Payne-Gaposchkin, "The Dyer's Hand," 101.

14. Payne-Gaposchkin, "The Dyer's Hand," 99–100.

15. John Robinson, *The Attwood Family: With Historic Notes and Pedigrees* (Sunderland: Hills and Company, 1903), 127.

16. Payne-Gaposchkin, "The Dyer's Hand," 81.

17. Payne-Gaposchkin, "The Dyer's Hand," 82.

18. Payne-Gaposchkin, "The Dyer's Hand," 105.

19. Payne-Gaposchkin, "The Dyer's Hand," 106–107.

20. Payne-Gaposchkin, "The Dyer's Hand," 106.

21. Payne-Gaposchkin, "The Dyer's Hand," 102.

22. Payne-Gaposchkin, "The Dyer's Hand," 93.

23. Payne-Gaposchkin, "The Dyer's Hand," 102.

24. Katharine Chorley, *Manchester Made Them* (London: Faber and Faber, 1950), 202.

25. Payne-Gaposchkin, "The Dyer's Hand," 109.

26. Payne-Gaposchkin, "The Dyer's Hand," 101.

27. Payne-Gaposchkin, "The Dyer's Hand," 102.

28. Payne-Gaposchkin, "The Dyer's Hand," 98, 102.

Chapter 4

1. Quoted in Jan Morris, *Oxford* (London: Faber and Faber, 1965), 99.

2. Henry Maudsley, "Sex in Mind and in Education," *Popular Science Monthly* 5 (June 1874): 199.

3. Carol Dyhouse, *Girls Growing Up in Late Victorian and Edwardian England* (London: Routledge, 1981), 153.

4. Howard Bailes, *Once a Paulina . . . : A History of St. Paul's Girls' School* (London: James & James, 2000), 68, 72.

5. Frances Gray to Harlow Shapley, January 13, 1923, UAV 630.22, box 7, folder 4, Harvard College Observatory, Records of Director Harlow Shapley, 1921–1956, Harvard University Archives.

6. Cecilia Payne-Gaposchkin, "The Dyer's Hand: An Autobiography," in *Cecilia Payne-Gaposchkin: An Autobiography and Other Recollections*, ed. Katherine Haramundanis (Cambridge: Cambridge University Press, 1984), 108.

7. Payne-Gaposchkin, "The Dyer's Hand," 219.

8. Payne-Gaposchkin, "The Dyer's Hand," 219.

9. *The Paulina* (student newspaper of St. Paul's School), January 1919. St. Paul's Girls' School Archives, London.

10. *The Paulina*, March 1919.

11. *The Paulina*, July 1919.

12. *The Paulina*, December 1919.

13. From a collection of letters in the file SPGS Correspondence: pre-World War II, St. Paul's Girls' School Archives. Thanks to Howard Bailes for this reference.

14. Ralph Vaughan Williams, "Holst, Gustav Theodore (1874–1934)," in *Dictionary of National Biography, 1931–1940* (London: Oxford University Press, 1949), https://archive.org/stream/in.ernet.dli.2015.146832/2015.146832.The-Dictionary-Of-National-Biography-1931-1940_djvu.txt.

15. Payne-Gaposchkin, "The Dyer's Hand," 108.

16. Gray to Shapley, January 13, 1923.

17. Minutes of the Governors' Meeting, St. Paul's School, May 19, 1911, St. Paul's Girls' School, Archives.

18. Minutes of the Governors' Meeting, St. Paul's School, July 19, 1918, St. Paul's Girls' School, Archives.

19. Payne-Gaposchkin, "The Dyer's Hand," 109.

20. Payne-Gaposchkin, "The Dyer's Hand," 110.

21. Katherine Haramundanis, "Cecilia Payne-Gaposchkin," in *Biographical Encyclopedia of Astronomers*, ed. Thomas Hockey, Virginia Trimble, Thomas R. Williams, et al., vol. 2 (New York: Springer, 2007), 877.

22. Payne-Gaposchkin, "The Dyer's Hand," 109.

Chapter 5

1. Cecilia Payne-Gaposchkin, "The Dyer's Hand: An Autobiography," in *Cecilia Payne-Gaposchkin: An Autobiography and Other Recollections*, ed. Katherine Haramundanis (Cambridge: Cambridge University Press, 1984), 109.

2. Payne-Gaposchkin, "The Dyer's Hand," 109.

3. Ann Phillips, ed., *A Newnham Anthology* (Cambridge: Cambridge University Press for Newnham College, 1979), 79.

4. Payne-Gaposchkin, "The Dyer's Hand," 112.

5. Mary Agnes Hamilton, *Newnham: An Informal Biography* (London: Faber and Faber, 1936), 173.

6. Phillips, *A Newnham Anthology*, 174.

7. Hamilton, *Newnham*, 164.

8. Phillips, *A Newnham Anthology*, 34.

9. Phillips, *A Newnham Anthology*, 159.

10. Phillips, *A Newnham Anthology*, 140.

11. Phillips, *A Newnham Anthology*, 166, 136.

12. Phillips, *A Newnham Anthology*, 120.

13. Phillips, *A Newnham Anthology*, 138, 176.

14. Phillips, *A Newnham Anthology*, 141.

15. Phillips, *A Newnham Anthology*, 140.

16. Phillips, *A Newnham Anthology*, 163.

17. Betty Grierson Leaf, "In Authority, 1923, Cecilia Payne," *Thersites* [Newnham College magazine], June 2, 1923, College Archives, Newnham College, Cambridge. The essay is reprinted as the foreword to "The Dyer's Hand."

18. Leaf, "In Authority."

19. Leaf, "In Authority."

20. Florence Nightingale, "Cassandra," *Cassandra and Other Selections from Suggestions for Thought*, ed. Mary Poovey (New York: New York University Press, 1992), 229.

21. Phillips, *A Newnham Anthology*, 54–55.

22. Phillips, *A Newnham Anthology*, 144.

Chapter 6

1. "About the University," University of Cambridge website, https://www
.cam.ac.uk/about-the-university/history/early-records.

2. Gillian Sutherland, "'Nasty forward minxes': Cambridge and the Higher
Education of Women," in *Cambridge Contributions*, ed. Sarah J. Ormrod (Cambridge:
Cambridge University Press, 1998), 89.

3. Mary Agnes Hamilton, *Newnham: An Informal Biography* (London: Faber
and Faber, 1936), 32.

4. Cecilia Payne-Gaposchkin, "The Dyer's Hand: An Autobiography," in *Cecilia
Payne-Gaposchkin: An Autobiography and Other Recollections*, ed. Katherine Haramundanis (Cambridge: Cambridge University Press, 1984), 112.

5. Payne-Gaposchkin, "The Dyer's Hand," 112.

6. Payne-Gaposchkin, "The Dyer's Hand," 113.

7. Stephen Jay Gould, *The Structure of Evolutionary Theory* (Cambridge, MA:
Harvard University Press, 2002), 396; "William Bateson, 1861–1926," *Journal of
Heredity* 17, no. 12 (1926): 445, 440.

8. *Complete Dictionary of Scientific Biography*, s.v. "William Bateson" (Scribner,
2008), https://www.encyclopedia.com/people/history/historians-miscellaneous
-biographies/william-bateson.

9. Payne-Gaposchkin, "The Dyer's Hand," 113.

10. Ann Phillips, ed., *A Newnham Anthology* (Cambridge: Cambridge University Press for Newnham College, 1979), 136.

11. Phillips, *A Newnham Anthology*, 125.

12. Hamilton, *Newnham*, 166; Phillips, *A Newnham Anthology*, 88.

13. Phillips, *A Newnham Anthology*, 125.

14. Phillips, *A Newnham Anthology*, 145.

15. Phillips, *A Newnham Anthology*, 111.

16. Payne-Gaposchkin, "The Dyer's Hand," 117.

Chapter 7

1. David Ball, "Cottingham, Edwin Turner (1869–1940), Modern Times," Ringstead People, http://ringstead.squarespace.com/ringstead-people/2010/10/14/cotting
ham-edwin-turner-1869-1940-modern-times.html; David Ball, "Edwin T. Cottingham
at the Science Museum," Ringstead People, http://ringstead.squarespace.com/ring
stead-people/2014/5/24/edwin-t-cottingham-at-the-science-museum.html; "Obituary
Notices: Fellows," *Monthly Notices of the Royal Astronomical Society*, 101 (1941): 131.

2. Quoted in Ball, "Cottingham, Edwin Turner (1869–1940)."

3. Ball, "Cottingham, Edwin Turner (1869–1940)."

4. Arthur Eddington, *Philosophy of Physical Science* (Cambridge: Cambridge University Press, 1939), 170, 171.

5. Lizzie Buchen, "May 29, 1919: A Major Eclipse, Relatively Speaking," *Wired*, May 29, 2009, emphasis in original.

6. Daniel Kennefick, "Testing Relativity from the 1919 Eclipse—A Question of Bias," *Physics Today* 62, no. 3 (2009): 38.

7. Subrahmanyan Chandrasekhar, "Verifying the Theory of Relativity," *Bulletin of the Atomic Scientists* 31, no. 6 (1975): 19.

8. David Bodanis, *Einstein's Greatest Mistake: A Biography* (Boston: Houghton Mifflin Harcourt, 2016), 98.

9. Matthew Stanley, "'An Expedition to Heal the Wounds of War': The 1919 Eclipse and Eddington as Quaker Adventurer," *Isis* 94, no. 1 (2003): 75.

10. Quoted in Bodanis, *Einstein's Greatest Mistake,* 100.

11. Quoted in Chandrasekhar, "Verifying the Theory of Relativity," 19.

12. Bodanis, *Einstein's Greatest Mistake,* 107.

13. Arthur Eddington to Albert Einstein, December 1, 1919, box 10, folder 2, Albert Einstein Papers, Department of Rare Books and Special Collections, Princeton University Library.

14. Cecilia Payne-Gaposchkin, "The Dyer's Hand: An Autobiography," in *Cecilia Payne-Gaposchkin: An Autobiography and Other Recollections,* ed. Katherine Haramundanis (Cambridge: Cambridge University Press, 1984), 117, 118.

15. On Arber, see Rudolf Schmid, "Agnes Arber, née Robertson (1879–1960): Fragments of Her Life, Including Her Place in Biology and in Women's Studies," *Annals of Botany* 88 (2001): 1105–1128.

16. Payne-Gaposchkin, "The Dyer's Hand," 114.

17. Payne-Gaposchkin, "The Dyer's Hand," 114.

18. Payne-Gaposchkin, "The Dyer's Hand," 113.

19. Payne-Gaposchkin, "The Dyer's Hand," 115.

20. Payne-Gaposchkin, "The Dyer's Hand," 112, 115.

Chapter 8

1. Cecilia Payne-Gaposchkin, "The Dyer's Hand: An Autobiography," in *Cecilia Payne-Gaposchkin: An Autobiography and Other Recollections,* ed. Katherine Haramundanis (Cambridge: Cambridge University Press, 1984), 118.

2. Ann Phillips, ed., *A Newnham Anthology* (Cambridge: Cambridge University Press for Newnham College, 1979), 159.

3. Betty Grierson Leaf, "In Authority, 1923, Cecilia Payne," *Thersites* [Newnham College magazine], June 2, 1923, College Archives, Newnham College, Cambridge. The essay is reprinted as the foreword to "The Dyer's Hand."

4. Phillips, *A Newnham Anthology*, 44, 156.

5. Virginia Trimble, correspondence with the author, October 2018. Trimble received a master's degree from the University of Cambridge in 1969.

6. Phillips, *A Newnham Anthology*, 156.

7. Phillips, *A Newnham Anthology*, 162, 157, 74, 139, 166.

8. Phillips, *A Newnham Anthology*, 166.

9. Phillips, *A Newnham Anthology*, 162.

10. Phillips, *A Newnham Anthology*, 84.

11. Quoted in Dennis Moralee, "The Old Cavendish—'The First Ten Years,'" in *A Hundred Years and More of Cambridge Physics*, ed. D. Moralee (Cambridge University Physics Society, 1974), https://www.phy.cam.ac.uk/history/old.

12. For more on the original Cavendish Lab, see Simon J. Schaffer, "A Tour Round the Old Cavendish Laboratory," https://www.youtube.com/watch?v=o7yNYF sglHE.

13. Schaffer, "A Tour Round the Old Cavendish Laboratory."

14. Quoted in Moralee, "The Old Cavendish—'The First Ten Years.'"

15. S. Devons, "Rutherford's Laboratory," in *A Hundred Years and More of Cambridge Physics*, ed. Moralee, https://www.phy.cam.ac.uk/history/years/rutherford.

16. Mark David Hurn, Departmental Librarian, Institute of Astronomy, University of Cambridge, interview with the author, May 2016.

17. Payne-Gaposchkin, "The Dyer's Hand," 116.

18. Payne-Gaposchkin, "The Dyer's Hand," 117.

19. John Heilbron, quoted in Tom Siegfried, "When the Atom Went Quantum," *ScienceNews*, July 13, 2013, 22.

20. Alfred Romer, "Proton or Prouton? Rutherford and the Depths of the Atom," *American Journal of Physics* 65, no. 8 (1997): 707.

21. Siegfried, "When the Atom Went Quantum," 20.

22. David H. DeVorkin, *Henry Norris Russell: Dean of American Astronomers* (Princeton, NJ: Princeton University Press, 2000), 219.

23. Siegfried, "When the Atom Went Quantum," 24.

24. Heilbron, quoted in Siegfried, "When the Atom Went Quantum."

25. Payne-Gaposchkin, "The Dyer's Hand," 116.

26. Payne-Gaposchkin, "The Dyer's Hand," 116.

Chapter 9

1. Cecilia Payne-Gaposchkin, "The Dyer's Hand: An Autobiography," in *Cecilia Payne-Gaposchkin: An Autobiography and Other Recollections*, ed. Katherine Haramundanis (Cambridge: Cambridge University Press, 1984), 118.

2. Payne-Gaposchkin, "The Dyer's Hand," 119.

3. Ann Phillips, ed., *A Newnham Anthology* (Cambridge: Cambridge University Press for Newnham College, 1979), 121.

4. Phillips, *A Newnham Anthology*, 122.

5. "The Equatorial Telescope," September 1914, p. 2, NHI/9, Newnham College Observatory, College Archives, Newnham College, Cambridge.

6. It is unclear from Cecilia's autobiography whether her first exposure to a telescope was at the Newnham Observatory or the Cambridge Observatory (see Chapter 10).

7. "Leslie Comrie," biographical sketch, Collegians at War, First World War Centenary, 2014–2018, Special Collections, University of Auckland, http://www.specialcollections.auckland.ac.nz/wwi-centenary/collegians-at-war/their-stories/leslie-comrie.

8. Donald H. Sadler, *A Personal History of H.M. Nautical Almanac Office (30 October 1930–18 February 1972)*, ed. George A. Wilkins (Taunton: UK Hydrographic Office, 2008), 33, http://astro.ukho.gov.uk/nao/history/dhs_gaw/nao_perhist_0802_dhs.pdf.

9. Interview of Cecilia Payne-Gaposchkin by Owen Gingerich, March 5, 1968, Niels Bohr Library and Archives, American Institute of Physics, www.aip.org/history-programs/niels-bohr-library/oral-histories/4620.

10. In her autobiography ("The Dyer's Hand," 121) Cecilia wrote that she fixed the telescope by herself; but in an interview with Owen Gingerich (cited above), she said that Comrie told her how to clean off the rust and "was very kind and helpful and came [to the observatory] several times."

11. Payne-Gaposchkin, "The Dyer's Hand," 121.

12. Betty Grierson Leaf, "In Authority, 1923, Cecilia Payne," *Thersites* [Newnham College magazine], June 2, 1923, College Archives, Newnham College, Cambridge. The essay is reprinted as the foreword to "The Dyer's Hand."

13. Leaf, "In Authority."

14. Ivan Leslie Thomsen, s.v. "Comrie, Leslie John, F.R.S.," *An Encyclopedia of New Zealand* (1966), https://teara.govt.nz/en/1966/comrie-leslie-john-frs.

15. Payne-Gaposchkin, "The Dyer's Hand," 122.

16. Phillips, *A Newnham Anthology*, 88.

17. Phillips, *A Newnham Anthology*, 134.

18. "Minutes of the Joint Committee of Staff and Students," Newnham College, November 1, 1921, St. Paul's Girls' School Archives, London.

19. Phillips, *A Newnham Anthology*, 47.

20. Phillips, *A Newnham Anthology*, 65, 148.

21. Phillips, *A Newnham Anthology*, 175.

22. Phillips, *A Newnham Anthology*, 112.

23. Phillips, *A Newnham Anthology*, 145.

24. Phillips, *A Newnham Anthology*, 108.

Chapter 10

1. Ann Phillips, ed., *A Newnham Anthology* (Cambridge: Cambridge University Press for Newnham College, 1979), 82, 103.

2. F. J. M. Stratton, *Annals of the Solar Physics Observatory, Cambridge* (Cambridge: Cambridge University Press, 1949), 5.

3. Stratton, *Annals of the Solar Physics Observatory*, 7.

4. D. W. Dewhirst, "A Note on Polar Refractors," *Journal for the History of Astronomy* 13 (1982): 119.

5. Cecilia writes of this visit in Cecilia Payne-Gaposchkin, "The Dyer's Hand: An Autobiography," in *Cecilia Payne-Gaposchkin: An Autobiography and Other Recollections*, ed. Katherine Haramundanis (Cambridge: Cambridge University Press, 1984), 119–120. All quotations are from these pages.

6. W. M. Smart, "Obituary Notices," *Monthly Notices of the Royal Astronomical Society*, 105 (1945): 91.

7. Payne-Gaposchkin, "The Dyer's Hand," 236.

8. Payne-Gaposchkin, "The Dyer's Hand," 120.

9. Payne-Gaposchkin, "The Dyer's Hand," 120.

10. Cecilia's descriptions of these men and her interactions with them are in Payne-Gaposchkin, "The Dyer's Hand," 120–121. All quotations are from these pages.

11. Payne-Gaposchkin, "The Dyer's Hand," 122.

12. Payne-Gaposchkin, "The Dyer's Hand," 122.

13. Payne-Gaposchkin, "The Dyer's Hand," 123.

14. Betty Grierson Leaf, "In Authority, 1923, Cecilia Payne," *Thersites* [Newnham College magazine], June 2, 1923, College Archives, Newnham College, Cambridge. The essay is reprinted as the foreword to "The Dyer's Hand."

15. Payne-Gaposchkin, "The Dyer's Hand," 123. The paper is: Cecilia H. Payne, "Proper Motions of the Stars in the Neighborhood of M36 (N.G.C. 1960)," *Monthly Notes of the Royal Astronomical Society* 83, no. 5 (1923): 334.

16. Payne-Gaposchkin, "The Dyer's Hand," 123.

17. William Wordsworth, *Lines Written a Few Miles above Tintern Abbey, On Revisiting the Banks of the Wye during a Tour* (1798), quoted in Payne-Gaposchkin, "The Dyer's Hand," 238.

Chapter 11

1. Ann Phillips, ed., *A Newnham Anthology* (Cambridge: Cambridge University Press for Newnham College, 1979), 162.

2. Phillips, *A Newnham Anthology*, 162, 139.

3. Phillips, *A Newnham Anthology*, 74, 45.

4. Phillips, *A Newnham Anthology*, 69, 139.

5. Phillips, *A Newnham Anthology*, 175.

6. Phillips, *A Newnham Anthology*, 139.

7. Cecilia Payne-Gaposchkin, "The Dyer's Hand: An Autobiography," in *Cecilia Payne-Gaposchkin: An Autobiography and Other Recollections*, ed. Katherine Haramundanis (Cambridge: Cambridge University Press, 1984), 220.

8. Payne-Gaposchkin, "The Dyer's Hand," 115.

9. Quoted in George Paget Thomson, "J. J. and the Cavendish," in Dennis Moralee, ed., *A Hundred Years and More of Cambridge Physics* (Cambridge University Physics Society, 1974), https://www.phy.cam.ac.uk/history/years/jjandcav. The author is the son of J. J. Thomson.

10. George Johnstone Stoney, "Of the 'Electron,' or Atom of Electricity," *London, Edinburgh, and Dublin Philosophical Magazine and Journal of Science* 38, no. 233 (1894): 418–420.

11. Quoted in Paula Gould, "Women and the Culture of University Physics in Late Nineteenth-Century Cambridge," *British Journal for the History of Science* 30, no. 2 (1997): 127.

12. Phillips, *A Newnham Anthology*, 77.

13. Phillips, *A Newnham Anthology*, 133.

14. Phillips, *A Newnham Anthology*, 133.

15. Quoted in E. N. da C. Andrade, *Rutherford and the Nature of the Atom* (New York: Peter Smith, 1964), 111.

16. Arthur S. Eve, *Rutherford* (Cambridge: Cambridge Press, 1939), 27.

17. S. Devons, "Rutherford's Laboratory," in *A Hundred Years and More of Cambridge Physics*, ed. Moralee, https://www.phy.cam.ac.uk/history/years/rutherford.

18. Devons, "Rutherford's Laboratory."

19. Mark Oliphant, *Rutherford: Recollections of the Cambridge Days* (Amsterdam: Elsevier, 1972), 26.

20. Devons, "Rutherford's Laboratory."

21. Payne-Gaposchkin, "The Dyer's Hand," 118.

22. Payne-Gaposchkin, "The Dyer's Hand," 118.

Chapter 12

1. Ann Phillips, ed., *A Newnham Anthology* (Cambridge: Cambridge University Press for Newnham College, 1979), 113.

2. Mary Agnes Hamilton, *Newnham: An Informal Biography* (London: Faber and Faber, 1936), 179. During Cecilia's years at Cambridge, the Senate was the university's governing body. It consisted of all holders of the MA degree and above. See University of Cambridge, "About the University," https://www.cam.ac.uk/about-the-university/how-the-university-and-colleges-work/governance.

3. Phillips, *A Newnham Anthology*, 150.

4. Phillips, *A Newnham Anthology*, 137.

5. "No Caps and Gowns for Women (1921)," newsreel footage, British Pathé, https://www.youtube.com/watch?v=d69I4iQ_Jow.

6. Phillips, *A Newnham Anthology*, 150.

7. Phillips, *A Newnham Anthology*, 150, 145.

8. Pat Thane, "The Careers of Female Graduates of Cambridge University, 1920s–1970s," in *Origins of the Modern Career*, ed. David Mitch, John Brown, and Marco H. D. van Leewen (Aldershot: Ashgate, 2004), 212, 213.

9. Phillips, *A Newnham Anthology*, 118.

10. Phillips, *A Newnham Anthology*, 133.

11. Cecilia Payne-Gaposchkin, "The Dyer's Hand: An Autobiography," in *Cecilia Payne-Gaposchkin: An Autobiography and Other Recollections*, ed. Katherine Haramundanis (Cambridge: Cambridge University Press, 1984), 124.

12. Carol Dyhouse, *Girls Growing Up in Late Victorian and Edwardian England* (London: Routledge, 1981), 73.

13. Payne-Gaposchkin, "The Dyer's Hand," 124.

14. Payne-Gaposchkin, "The Dyer's Hand," 124.

15. Payne-Gaposchkin, "The Dyer's Hand," 158.

16. Payne-Gaposchkin, "The Dyer's Hand," 124.

17. Payne-Gaposchkin, "The Dyer's Hand," 124.

18. Phillips, *A Newnham Anthology*, 95, 109.

19. Phillips, *A Newnham Anthology*, 120.

20. Phillips, *A Newnham Anthology*, 109.

21. Payne-Gaposchkin, "The Dyer's Hand," 118–119.

22. Payne-Gaposchkin, "The Dyer's Hand," 220.

23. Betty Grierson Leaf, "In Authority, 1923, Cecilia Payne," *Thersites* [Newnham College magazine], June 2, 1923, College Archives, Newnham College, Cambridge. The essay is reprinted as the foreword to "The Dyer's Hand."

24. Payne-Gaposchkin, "The Dyer's Hand," 220, 125.

Chapter 13

1. Letter from Eleanor Mildred Sidgwick, January 1905, EC/2/4/7, College Archives, Newnham College.

2. Cecilia Payne to Harlow Shapley, February 26, 1923, UAV 630.22, box 7, folder 4, Harvard College Observatory, Records of Director Harlow Shapley, 1921–1956, Harvard University Archives (hereafter, Shapley Records).

3. Arthur Eddington to Harlow Shapley, January 18, 1923, UAV 630.22, box 7, folder 4. Shapley Records.

4. L. J. Comrie to Harlow Shapley, March 7, 26, 1923, UAV 630.22, box 4, folder 3, Shapley Records.

5. Frances Gray to Harlow Shapley, January 13, 1923, UAV 630.22, box 7, folder 4, Shapley Records.

6. G. F. C. Searle to Harlow Shapley, January 16, 1923, UAV 630.22, box 7, folder 4, Shapley Records.

7. Harlow Shapley to Cecilia Payne, March 12, 1923, UAV 630.22, box 7, folder 4, Shapley Records.

8. Cecilia Payne to Harlow Shapley, April 5, 1923, UAV 630.22, box 7, folder 4, Shapley Records.

9. Harlow Shapley to Cecilia Payne, April 16, 1923, UAV 630.22, box 7, folder 4, Shapley Records. Cecilia was eventually awarded a Pickering Fellowship. See A. J. Cannon, "Report of the Astronomical Fellowship Committee," *Annual Report of the Maria Mitchell Association* 25 (1927): 14.

10. Cecilia Payne to Harlow Shapley, April 27, 1923, UAV 630.22, box 7, folder 4, Shapley Records.

11. Cecilia Payne-Gaposchkin, "The Dyer's Hand: An Autobiography," in *Cecilia Payne-Gaposchkin: An Autobiography and Other Recollections*, ed. Katherine Haramundanis (Cambridge: Cambridge University Press, 1984), 124.

12. Cecilia Payne to Harlow Shapley, June 23, 1923, UAV 630.22, box 7, folder 4, Shapley Records.

13. Payne-Gaposchkin, "The Dyer's Hand," 111.

14. Payne-Gaposchkin, "The Dyer's Hand," 125.

15. Peggy A. Kidwell, "An Historical Introduction to 'The Dyer's Hand,'" in *Cecilia Payne-Gaposchkin*, ed. Haramundanis, 12.

16. Interview of Cecilia Payne-Gaposchkin by Owen Gingerich, March 5, 1968, Niels Bohr Library and Archives, American Institute of Physics, www.aip.org/history-programs/niels-bohr-library/oral-histories/4620.

17. Payne-Gaposchkin, "The Dyer's Hand," 155.

18. Carol Dyhouse, *Girls Growing Up in Late Victorian and Edwardian England* (London: Routledge, 1981), 74.

Chapter 14

1. Cecilia Payne-Gaposchkin, "The Dyer's Hand: An Autobiography," in *Cecilia Payne-Gaposchkin: An Autobiography and Other Recollections*, ed. Katherine Haramundanis (Cambridge: Cambridge University Press, 1984), 129.

2. Payne-Gaposchkin, "The Dyer's Hand," 129.

3. Payne-Gaposchkin, "The Dyer's Hand," 131.

4. Payne-Gaposchkin, "The Dyer's Hand," 221.

5. Payne-Gaposchkin, "The Dyer's Hand," 136.

6. Photograph in the Katherine Haramundanis collection.

7. Photograph in the Katherine Haramundanis collection.

8. Solon I. Bailey, *The History and Work of the Harvard Observatory, 1839 to 1927* (New York: McGraw-Hill, 1931), 23.

9. Bailey, *The History and Work of the Harvard Observatory*, 25.

10. Katherine Haramundanis, "A Personal Recollection," in *Cecilia Payne-Gaposchkin*, ed. Haramundanis, 47.

11. Bailey, *The History and Work of the Harvard Observatory*, 247.

12. Pamela Mack, "Straying from Their Orbits," in *Women of Science: Righting the Record*, ed. G. Kass-Simon and Patricia Farnes (Bloomington: Indiana University Press, 1990), 92.

13. Elizabeth Agassiz, Radcliffe Commencement Address, June 21, 1899, quoted in Bernice Brown Cronkhite, "Story of the Zemurray Professorship," 1968, Radcliffe College Archives subject files, Women faculty at Harvard, RG XXIV, ser. 5, folder 25.10, Schlesinger Library, Radcliffe Institute, Harvard University, https://iiif.lib.harvard.edu/manifests/view/drs:51697047$23i.

14. Harriet Richardson Donaghe, "Photographic Flashes from Harvard Observatory," *Popular Astronomy* 6, no. 9 (1898): 450.

15. Williamina Paton Stevens Fleming, "Journal of Williamina Paton Fleming," 1900, p. 7, Harvard University Archives, http://nrs.harvard.edu/urn-3:HUL.ARCH:666402.

16. E. Dorrit Hoffleit, "Pioneering Women in the Spectral Classification of Stars," *Physics in Perspective* 4 (2002): 383.

17. Hoffleit, "Pioneering Women," 384.

18. *Vassar Encyclopedia Online*, s.v. "Antonia Maury," http://vcencyclopedia.vassar.edu/alumni/antonia-maury.html.

19. Payne-Gaposchkin, "The Dyer's Hand," 141, 148, 149, 140.

20. Quoted in Mack, "Straying from Their Orbits," 96.

21. Mack, "Straying from Their Orbits," 100.

22. Margaret Mayall, interview held in Cambridge, MA, December 8, 1976, quoted in Mack, "Straying from Their Orbits," 100.

23. Sarah F. Whitney, draft of a biography of Annie Jump Cannon for *The Biographical Dictionary of American Women*, date unknown, HUGFP 125.95, box 1, Annie Jump Cannon Papers, Harvard Archives, Harvard University. See also Hoffleit, "Pioneering Women," 389.

24. Annie Jump Cannon, Diary, September 21, 1885, private collection of Margaret Mayall, quoted in Mack, "Straying from Their Orbits," 98.

25. Pickering, quoted in Hoffleit, "Pioneering Women," 389.

26. Payne-Gaposchkin, "The Dyer's Hand," 149.

27. Payne-Gaposchkin, "The Dyer's Hand," 139.

28. Edward Pickering, "Periods of 25 Variable Stars in the Small Magellanic Cloud," *Harvard College Observatory Circular* 173 (March 12, 1912).

29. Payne-Gaposchkin, "The Dyer's Hand," 149.

30. Payne-Gaposchkin, "The Dyer's Hand," 141, 143.

31. Payne-Gaposchkin, "The Dyer's Hand," 143.

32. Peggy A. Kidwell, "Cecilia Payne-Gaposchkin: Astronomy in the Family," in *Uneasy Careers and Intimate Lives*, ed. Pnina G. Abir-Am and Dorinda Outram (Rutgers: Rutgers University Press, 1987), 220.

33. Quoted in Mack, "Straying from Their Orbits," 104.

34. Payne-Gaposchkin, "The Dyer's Hand," 153.

Chapter 15

1. Cecilia Payne-Gaposchkin, "The Dyer's Hand: An Autobiography," in *Cecilia Payne-Gaposchkin: An Autobiography and Other Recollections*, ed. Katherine Haramundanis (Cambridge: Cambridge University Press, 1984), 137.

2. Payne-Gaposchkin, "The Dyer's Hand," 137.

3. Bart J. Bok, *Harlow Shapley, 1885–1972: A Biographical Memoir* (Washington, DC: National Academy of Sciences, 1978).

4. Quoted in Owen Gingerich, "How Shapley Came to Harvard, or, Snatching the Prize from the Jaws of Debate," *Journal for the History of Astronomy* 19, no. 3 (1988): 201.

5. Quoted in Gingerich, "How Shapley Came to Harvard," 202.

6. Quoted in Gingerich, "How Shapley Came to Harvard," 202, 203.

7. Quoted in Gingerich, "How Shapley Came to Harvard," 204.

8. Payne-Gaposchkin, "The Dyer's Hand," 155. Although Cecilia never mentions being granted a Pickering Fellowship, Annie Cannon, writing in the observatory's

"Report of the Astronomical Fellowship Committee," lists Cecilia as receiving a Pickering Fellowship grant in 1923–1924: A. J. Cannon, "Report of the Astronomical Fellowship Committee," *Annual Report of the Maria Mitchell Association* 25 (1927): 14.

9. Interview of Cecilia Payne-Gaposchkin by Owen Gingerich, March 5, 1968, Niels Bohr Library and Archives, American Institute of Physics, www.aip.org/history-programs/niels-bohr-library/oral-histories/4620.

10. Payne-Gaposchkin, "The Dyer's Hand," 157.

11. Payne-Gaposchkin, "The Dyer's Hand," 157.

12. Payne-Gaposchkin, "The Dyer's Hand," 160.

13. Payne-Gaposchkin, "The Dyer's Hand," 190.

14. Payne-Gaposchkin, "The Dyer's Hand," 182.

15. Katherine Gaposchkin Haramundanis, "Cecilia Payne-Gaposchkin: A Stellar Pioneer," Dorrit Hoffleit Lecture, American Physical Society, April 23, 2006, reprinted in *CSWP Gazette* 25, no. 2 (2006): 6.

16. Payne-Gaposchkin, "The Dyer's Hand," 150, 139.

17. Payne-Gaposchkin, "The Dyer's Hand," 139.

18. "Seeking Body of Adelaide Ames," *Boston Herald*, June 28, 1932.

19. *Boston Transcript*, June 27, 1932. Quoted in "Research Astronomer Lost by Drowning," *Popular Astronomy* 40 (1932): 448.

20. Payne-Gaposchkin, "The Dyer's Hand," 142, 189.

21. Payne-Gaposchkin, "The Dyer's Hand," 154.

22. Payne-Gaposchkin, "The Dyer's Hand," 154, 155.

23. Payne-Gaposchkin, "The Dyer's Hand," 156.

24. Payne-Gaposchkin, "The Dyer's Hand," 222.

25. Kay Redfield Jamison, *Exuberance: The Passion for Life* (New York: Knopf, 2004), 173.

26. Payne-Gaposchkin, "The Dyer's Hand," 159.

27. Katherine Haramundanis, "A Personal Recollection," in *Cecilia Payne-Gaposchkin*, ed. Haramundanis, 43.

28. Payne-Gaposchkin, "The Dyer's Hand," 150.

29. Payne-Gaposchkin, "The Dyer's Hand," 163.

30. Emma Payne to Harlow Shapley, March 22, 1924, UAV 630.22, box 12, Harvard College Observatory, Records of Director Harlow Shapley, 1921–1956, Harvard University Archives.

31. Payne-Gaposchkin, "The Dyer's Hand," 114.

32. Payne-Gaposchkin, "The Dyer's Hand," 159.

33. Payne-Gaposchkin, "The Dyer's Hand," 163.

34. Payne-Gaposchkin, "The Dyer's Hand," 220–221.

35. Payne-Gaposchkin, "The Dyer's Hand," 164.

36. Payne-Gaposchkin, "The Dyer's Hand," 165.

Chapter 16

1. Cecilia Payne-Gaposchkin, "The Dyer's Hand: An Autobiography," in *Cecilia Payne-Gaposchkin: An Autobiography and Other Recollections,* ed. Katherine Haramundanis (Cambridge: Cambridge University Press, 1984), 161.

2. Henry Norris Russell to Harlow Shapley, January 31, 1920, quoted in David H. DeVorkin, *Henry Norris Russell: Dean of American Astronomers* (Princeton, NJ: Princeton University Press, 2000), 169.

3. Payne-Gaposchkin, "The Dyer's Hand," 162.

4. Payne-Gaposchkin, "The Dyer's Hand," 162.

5. Henry Norris Russell to Harlow Shapley, October 30, 1923, quoted in DeVorkin, *Henry Norris Russell,* 202.

6. Kay Redfield Jamison, *Exuberance: The Passion for Life* (New York: Knopf, 2004), 192.

7. Interview of Cecilia Payne-Gaposchkin by Owen Gingerich, March 5, 1968, Niels Bohr Library and Archives, American Institute of Physics, https://www.aip.org/history-programs/niels-bohr-library/oral-histories/4620.

8. Payne-Gaposchkin, "The Dyer's Hand," 140.

9. Cecilia Payne-Gaposchkin, *Stars in the Making* (New York: Pocket Books, 1959), xi.

10. Quoted in D. S. Kothari, "Meghnad Saha," *Biographical Memoirs of Fellows of the Royal Society* 5 (1959): 221.

11. Kothari, "Meghnad Saha," 222; Meghnad Saha, "On a Physical Theory of Stellar Spectra," *Proceedings of the Royal Society of London* A99 (1921): 135–153.

12. David H. DeVorkin, "Quantum Physics and the Stars (IV): Meghnad Saha's Fate," *Journal for the History of Astronomy* 25 (1994): 158.

13. Quoted in David H. DeVorkin and Ralph Kenat, "Quantum Physics and the Stars (II): Henry Norris Russell and the Abundances of the Elements in the Atmospheres of the Sun and Stars," *Journal for the History of Astronomy* 14, no. 3 (1983): 180.

14. DeVorkin and Kenat, "Quantum Physics and the Stars (II)," 191, 192.

15. Payne-Gaposchkin, "The Dyer's Hand," 160.

16. Marcus Chown, "The Woman Who Dissected the Sun," *New Scientist,* November 8, 2003.

17. DeVorkin, *Henry Norris Russell,* 199.

18. Owen Gingerich, "The Most Brilliant Ph.D. Thesis Ever Written in Astronomy," in *The Starry Universe: The Cecilia Payne-Gaposchkin Centenary*, ed. A. G. Davis Philip and Rebecca A. Koopmann (Schenectady, NY: L. Davis Press, 2000), 4.

19. Payne-Gaposchkin, "The Dyer's Hand," 221.

20. Payne-Gaposchkin, "The Dyer's Hand," 177.

21. Payne-Gaposchkin, "The Dyer's Hand," 177.

22. Gingerich, "The Most Brilliant Ph.D. Thesis," 11–12.

23. Frank Schlesinger and Dirk Brouwer, "Biographical Memoir of Ernest William Brown, 1866–1938," *National Academy of Sciences Biographical Memoirs* 21 (1939): 259.

24. Cecilia Payne to Margaret Harwood, January 9, 1925, carton 2, folder 50, Margaret Harwood Papers, 1902–1974, 79-M62—2007-M228, Schlesinger Library, Radcliffe Institute, Harvard University.

25. Payne to Harwood, January 9, 1925.

26. Payne to Harwood, January 9, 1925.

27. Quoted in David H. DeVorkin, "Extraordinary Claims Require Extraordinary Evidence: C. H. Payne, H. N. Russell and Standards of Evidence in Early Quantitative Stellar Spectroscopy," *Journal of Astronomical History and Heritage* 13, no. 2 (2010): 141.

28. Cecilia Payne to Margaret Harwood, August 16, 1924, carton 2, folder 50, Margaret Harwood Papers, 1902–1974, 79-M62—2007-M228, Schlesinger Library, Radcliffe Institute, Harvard University.

29. Payne-Gaposchkin, "The Dyer's Hand," 156. Cecilia is referring to a sixteenth-century illustration and ballad, a metaphor for love at first sight employed occasionally by Shakespeare. Cophetua was an African king who was not particularly attractive to women. When he happened to spot a beggar from his palace window one day, he asked her to marry him; she agreed and they ended up living a quiet, happy life.

30. Payne-Gaposchkin, "The Dyer's Hand," 157.

31. P. W. Bridgman, "Theodore Lyman, 1874–1954: A Biographical Memoir" (Washington, DC: National Academy of Sciences, 1957).

32. Harlow Shapley to Theodore Lyman, September 11, 1924, UAV 630.22, box 12, Harvard College Observatory, Records of Director Harlow Shapley, 1921–1956, Harvard University Archives (hereafter Shapley Records).

33. Theodore Lyman to Harlow Shapley, September 20, 1924, UAV 630.22, box 12, Shapley Records.

34. Payne-Gaposchkin, "The Dyer's Hand," 157.

35. Henry Norris Russell to Harlow Shapley, August 11, 1925, quoted in Gingerich, "The Most Brilliant Ph.D. Thesis," 9.

36. DeVorkin and Kenat, "Quantum Physics and the Stars (II)," 181.

37. Owen Gingerich, "A Star Is Born," review of *Henry Norris Russell: Dean of American Astronomers*, by Daniel A. DeVorkin, *New York Times*, November 12, 2000.

38. Henry Norris Russell to Cecilia Payne, January 14, 1925, quoted in DeVorkin and Kenat, "Quantum Physics and the Stars (II)," 187.

39. Payne-Gaposchkin, "The Dyer's Hand," 177.

40. Cecilia Payne-Gaposchkin to Charlotte Moore Sitterly, February 24, 1977, Papers of Cecilia Helena Payne-Gaposchkin, 1924, circa 1950s–1990s, 2000, HUGBP 182.50, Harvard University Archives.

41. Payne to Harwood, January 9, 1925.

42. Marcia Bartusiak, "The Stuff of Stars," *The Sciences* (September/October 1993): 37.

43. Cecilia H. Payne, *Stellar Atmospheres: A Contribution to the Observational Study of High Temperature in the Reversing Layers of Stars*, Harvard Observatory Monographs, ed. Harlow Shapley (Cambridge, MA: Harvard College Observatory, 1925), 186, 188.

44. DeVorkin, *Henry Norris Russell*, 204.

45. DeVorkin, "Extraordinary Claims Require Extraordinary Evidence,"139.

46. Cecilia Payne to Harlow Shapley, June 10, 1925, UAV 630.22, box 7, folder 4, Shapley Records.

47. Gingerich, "The Most Brilliant Ph.D. Thesis," 8.

48. Harlow Shapley to Henry Norris Russell, August 6, 1925; Harlow Shapley to Ada Comstock, September 14, quoted in Peggy A. Kidwell, "An Historical Introduction to 'The Dyer's Hand,'" in *Cecilia Payne-Gaposchkin*, ed. Haramundanis, 20.

49. Interview with Katherine Haramundanis, May 2017.

50. Quoted in Bartusiak, "The Stuff of Stars," 37.

Chapter 17

1. Dorrit Hoffleit, "Reminiscences of Cecilia Payne-Gaposchkin (1900–1979)," in *The Starry Universe: The Cecilia Payne-Gaposchkin Centenary*, ed. A. G. Davis Philip and Rebecca A. Koopmann (Schenectady, NY: L. Davis Press, 2000), 91.

2. Cecilia Payne-Gaposchkin, "The Dyer's Hand: An Autobiography," in *Cecilia Payne-Gaposchkin: An Autobiography and Other Recollections*, ed. Katherine Haramundanis (Cambridge: Cambridge University Press, 1984), 167.

3. Lady Bertha Swirles Jeffreys to Katherine Haramundanis, August 11, 1985, quoted in Katherine Gaposchkin Haramundanis, "Cecilia and Her World," in *The Starry Universe*, ed. Philip and Koopmann, 20.

4. Payne-Gaposchkin, "The Dyer's Hand," 165.

5. Payne-Gaposchkin, "The Dyer's Hand," 133.

6. Cecilia Payne to Harlow Shapley, August 10, 1924, UAV 630.22, box 12, Harvard College Observatory, Records of Director Harlow Shapley, 1921–1956, Harvard University Archives (hereafter Shapley Records).

7. Payne-Gaposchkin, "The Dyer's Hand," 166.

8. Payne-Gaposchkin, "The Dyer's Hand," 133, 167.

9. Payne-Gaposchkin, "The Dyer's Hand," 171.

10. Frances Wright to Harlow Shapley, March 29, 1929, HUG 4773.10, box 105, Papers of Harlow Shapley, 1906–1966, Harvard University Archives (hereafter Shapley Papers).

11. Frances Wright to Harlow Shapley, February 19, 1930, box 105, Shapley Papers.

12. Cecilia Payne-Gaposchkin, "No Wine So Wonderful as Thirst," *Radcliffe Quarterly* 41, no. 1 (1957): 11.

13. Payne-Gaposchkin, "The Dyer's Hand," 167.

14. David H. DeVorkin and Ralph Kenat, "Quantum Physics and the Stars (II): Henry Norris Russell and the Abundances of the Elements in the Atmospheres of the Sun and Stars," *Journal for the History of Astronomy* 14, no. 3 (1983): 206.

15. Arthur S. Eddington, *The Internal Constitution of the Stars* (Cambridge: Cambridge University Press, 1926), 369.

16. Payne-Gaposchkin, "The Dyer's Hand," 184.

17. Peggy A. Kidwell, "An Historical Introduction to 'The Dyer's Hand,'" in *Cecilia Payne-Gaposchkin*, ed. Haramundanis, 25.

18. J. S. Paraskevopoulos to Harlow Shapley, May 30, 1933, quoted in Kidwell, "An Historical Introduction," 26.

19. Cecilia Payne to Margaret Harwood, November 22, 1924, carton 2, folder 50, Margaret Harwood Papers, 1902–1974, 79-M62—2007-M228, Schlesinger Library, Radcliffe Institute, Harvard University.

20. Payne-Gaposchkin, "The Dyer's Hand," 220.

21. Interview of Nan Dieter-Conklin by David DeVorkin, July 19, 1977, Niels Bohr Library and Archives, American Institute of Physics, https://www.aip.org/history-programs/niels-bohr-library/oral-histories/4573.

22. Jesse L. Greenstein, "An Introduction to 'The Dyer's Hand,'" in *Cecilia Payne-Gaposchkin*, ed. Haramundanis, 8.

23. Owen Gingerich, "The Most Brilliant Ph.D. Thesis Ever Written in Astronomy," in *The Starry Universe*, ed. Philip and Koopmann, 3.

24. Harlow Shapley to the Harvard University Bursar, May 4, 1929, and March 24, 1930, quoted in Kidwell, "An Historical Introduction," 26.

25. Jerome Karabel, *The Chosen: The Hidden History of Admission and Exclusion at Harvard, Yale, and Princeton* (Boston: Houghton Mifflin, 2005), 101.

26. Payne-Gaposchkin, "The Dyer's Hand," 221.

27. Harlow Shapley to Edward Pendray, October 22, 1934, HUG 4773.10, box 117, "Literary Digest" file folder, Shapley Papers.

28. Henry Norris Russell to J. S. Plaskett, November 4, 1924, quoted in Kidwell, "An Historical Introduction," 25.

29. J. S. Plaskett to Henry Norris Russell, January 8, 1925, quoted in Kidwell, "An Historical Introduction," 25.

30. Cecilia Payne to Henry Norris Russell, December 11, 1930, quoted in Kidwell, "An Historical Introduction," 26.

31. Kidwell, "An Historical Introduction," 27; Payne-Gaposchkin, "The Dyer's Hand," 222.

32. Payne-Gaposchkin, "The Dyer's Hand," 183.

33. Payne-Gaposchkin, "The Dyer's Hand," 133.

34. Payne-Gaposchkin, "The Dyer's Hand," 143.

35. Payne-Gaposchkin, "The Dyer's Hand," 183.

36. Payne-Gaposchkin, "The Dyer's Hand," 184.

37. Quoted in DeVorkin and Kenat, "Quantum Physics and the Stars (II)," 207.

38. David H. DeVorkin, *Henry Norris Russell: Dean of American Astronomers* (Princeton, NJ: Princeton University Press, 2000), 199. See also Henry Norris Russell, "On the Composition of the Sun's Atmosphere," *Astrophysical Journal* 70 (1929): 79.

39. DeVorkin and Kenat, "Quantum Physics and the Stars (II)," 208.

40. Payne-Gaposchkin, "The Dyer's Hand," 172.

41. Cecilia Payne-Gaposchkin to Harlow Shapley, box 12, Shapley Records, letters dated August 23, 1924; July 28, 1925; n.d., 1927; December 7, 1928.

42. Hoffleit, "Reminiscences of Cecilia Payne-Gaposchkin," 91.

43. Katherine Haramundanis, "A Personal Recollection," in *Cecilia Payne-Gaposchkin*, ed. Haramundanis, 44.

44. Norbert Wiener, *Ex-Prodigy: My Childhood and Youth* (Cambridge, MA: MIT Press, 1953), 67.

45. *The World Magazine*, October 7, 1906, 1.

46. Larry Hardesty, "The Original Absent-Minded Professor," *MIT Technology Review*, June 21, 2011.

47. Flo Conway and Jim Siegelman, *Dark Hero of the Information Age: In Search of Norbert Wiener, the Father of Cybernetics* (New York, NY: Basic Books, 2005), 59.

48. Norbert Wiener to Constance, July 5, 1925, MC-0022, box X, folder X, Norbert Wiener Papers, Department of Distinctive Collections, MIT, Cambridge, MA (hereafter Wiener Papers).

49. Norbert Wiener to Fritz, June 24, 1925, Wiener Papers.

50. Norbert Wiener to his parents, June 1925, Wiener Papers.

51. Norbert Wiener to Constance, August 27, 1925, Wiener Papers.

52. Norbert Wiener to Constance, July 5, 1925, Wiener Papers.

53. Norbert Wiener to Constance, August 27, 1925, Wiener Papers.

54. Norbert Wiener to his mother, September 1, 1925, Wiener Papers.

55. Carol Dyhouse, *Girls Growing Up in Late Victorian and Edwardian England* (London: Routledge, 1981), 44.

56. Norbert Wiener to Constance, July 21, 1925, Wiener Papers.

57. Norbert Wiener to Fritz, July 21, 1925, Wiener Papers.

58. Norbert Wiener to Fritz, March 4, 1926, Wiener Papers.

59. Conway and Siegelman, *Dark Hero of the Information Age*, 62.

60. Russell, "On the Composition of the Sun's Atmosphere," 22.

61. Russell, "On the Composition of the Sun's Atmosphere," 79.

62. R. d'E. Atkinson, "Atomic Synthesis and Stellar Energy," *Astrophysical Journal* 73 (1931): 254.

63. Russell, "On the Composition of the Sun's Atmosphere," 65.

64. DeVorkin and Kenat, "Quantum Physics and the Stars (II)," 216.

65. Owen Gingerich, "A Star Is Born," review of *Henry Norris Russell: Dean of American Astronomers*, by Daniel A. DeVorkin, *New York Times*, November 12, 2000.

66. DeVorkin, *Henry Norris Russell*, 366.

67. Gingerich, "The Most Brilliant Ph.D. Thesis," 14.

68. Quoted in DeVorkin, *Henry Norris Russell*, 341.

69. Payne-Gaposchkin, "The Dyer's Hand," 222.

70. Payne-Gaposchkin, "The Dyer's Hand," 223.

71. Payne-Gaposchkin, "The Dyer's Hand," 221.

72. Cecilia Payne-Gaposchkin, *Introduction to Astronomy* (Englewood Cliffs, NJ: Prentice-Hall, Inc., 1954).

73. Payne-Gaposchkin, "The Dyer's Hand," 175.

74. Payne-Gaposchkin, *Introduction to Astronomy*, 266.

75. Janet Akyüz Mattei and Kerriann H. Malatesta, "New Directions in Variable Star Research," in *The Starry Universe*, ed. Philip and Koopmann, 67.

76. Payne-Gaposchkin, "The Dyer's Hand," 223.

77. Payne-Gaposchkin, "The Dyer's Hand," 223.

78. Payne-Gaposchkin, "The Dyer's Hand," 223–224.

79. Payne-Gaposchkin, "The Dyer's Hand," 224.

80. Payne-Gaposchkin, "The Dyer's Hand," 224.

81. *Boston Herald*, June 28, 1932.

82. Payne-Gaposchkin, "The Dyer's Hand," 189–190.

83. Payne-Gaposchkin, "The Dyer's Hand," 190–191.

84. Payne-Gaposchkin, "The Dyer's Hand," 191.

Chapter 18

1. For biographical information on Sergei Gaposchkin, see *The Biographical Encyclopedia of Astronomers*, s.v. "Gaposchkin, Sergei [Sergej] Illarionovich," by K. Haramundanis (New York: Springer, 2007). For more on Sergei's life, drawn mostly from his unpublished autobiography, see Sylvia L. Boyd, *Portrait of a Binary: The Lives of Cecilia Payne and Sergei Gaposchkin* ([Rockland, ME]: Penobscot Press, 2014).

2. Cecilia Payne-Gaposchkin, "The Dyer's Hand: An Autobiography," in *Cecilia Payne-Gaposchkin: An Autobiography and Other Recollections*, ed. Katherine Haramundanis (Cambridge: Cambridge University Press, 1984), 191.

3. Simon Werrett, "The Astronomical Capital of the World: Pulkovo Observatory in the Russia of Tsar Nicholas I," in *The Heavens on Earth: Observatories and Astronomy in Nineteenth-Century Science and Culture*, ed. David Aubin, Charlotte Bigg, and Otto H. Sibum (Durham: Duke University Press, 2010), 33–57.

4. Eufrosina Dvoichenko-Markov, "The Pulkovo Observatory and Some American Astronomers of the Mid-19th Century," *Isis* 43, no. 3 (1952); Payne-Gaposchkin, "The Dyer's Hand," 190.

5. This description of Payne-Gaposchkin's trip is from "The Dyer's Hand," 192–197. All quotations are from these pages unless otherwise indicated.

6. Boris Gerasimovič to Harlow Shapley, August 18, 1933, UAV 630.22, box 33, Harvard College Observatory, Records of Director Harlow Shapley, 1921–1956, Harvard University Archives (hereafter Shapley Records).

7. *The Biographical Encyclopedia of Astronomers*, s.v. "Gerasimovich [Gerasimovič], Boris Petrovich," by K. Haramundanis (New York: Springer, 2007); Robert A. McCutcheon, "The 1936–1937 Purge of Soviet Astronomers," *Slavic Review* 50, no. 1 (1991): 100–117.

8. Sergei Gaposchkin, "Life (addition to thesis 1932)," UAV 630.22, box 33, Shapley Records.

9. Payne-Gaposchkin, "The Dyer's Hand," 196–197.

10. Cecilia Payne-Gaposchkin to Harlow Shapley, August 12, 1933, UAV 630.22, box 33, Shapley Records.

11. Payne-Gaposchkin to Shapley, August 12, 1933.

12. Peggy A. Kidwell, "Cecilia Payne-Gaposchkin: Astronomy in the Family," in *Uneasy Careers and Intimate Lives*, ed. Pnina G. Abir-Am and Dorinda Outram (New Brunswick, NJ: Rutgers University Press, 1987), 227.

13. Harlow Shapley to Sergei Gaposchkin, September 1, 1933, UAV 630.22, box 33, Shapley Records.

14. Payne-Gaposchkin, "The Dyer's Hand," 197.

15. Harlow Shapley to Sergei Gaposchkin, October 10, 1933, UAV 630.22, box 33, Shapley Records.

16. Harlow Shapley to George S. Messersmith, October 10, 1933, UAV 630.22, box 33, Shapley Records.

17. Sergei Gaposchkin to Harlow Shapley, November 1933, UAV 630.22, box 33, Shapley Records.

18. Payne-Gaposchkin, "The Dyer's Hand," 197.

19. Cecilia Payne-Gaposchkin to Harlow Shapley, March 6, 1934, UAV 630.22, box 33, Shapley Records.

20. Emma Payne to Harlow Shapley, April 11, 1934, UAV 630.22, box 33, Shapley Records.

21. Payne-Gaposchkin to Shapley, March 6, 1934.

22. Henry Norris Russell to Joseph Boyce, March 15, 1934, quoted in Peggy A. Kidwell, "Cecilia Payne-Gaposchkin: Astronomy in the Family," in *Uneasy Careers and Intimate Lives*, ed. Pnina G. Abir-Am and Dorinda Outram (New Brunswick, NJ: Rutgers University Press, 1987), 228.

Chapter 19

1. Dorrit Hoffleit, "Reminiscences of Cecilia Payne-Gaposchkin (1900–1979)," in *The Starry Universe: The Cecilia Payne-Gaposchkin Centenary*, ed. A. G. Davis Philip and Rebecca A. Koopmann (Schenectady, NY: L. Davis Press, 2000), 94.

2. Cecilia Payne-Gaposchkin, "The Dyer's Hand: An Autobiography," in *Cecilia Payne-Gaposchkin: An Autobiography and Other Recollections*, ed. Katherine Haramundanis (Cambridge: Cambridge University Press, 1984), 199.

3. Janet Akyüz Mattei and Kerriann H. Malatesta, "New Directions in Variable Star Research," in *The Starry Universe*, ed. Philip and Koopmann, 69.

4. Payne-Gaposchkin, "The Dyer's Hand," 215.

5. Payne-Gaposchkin, "The Dyer's Hand," 198.

6. Owen Gingerich, "The Most Brilliant Ph.D. Thesis Ever Written in Astronomy," in *The Starry Universe*, ed. Philip and Koopmann, 5.

7. Payne-Gaposchkin, "The Dyer's Hand," 198.

8. Cecilia Payne-Gaposchkin and Sergei Gaposchkin, *Variable Stars* (Cambridge, MA: Harvard Observatory, 1938); Dorrit Hoffleit, "The Milton Bureau Revisited," *Journal of the American Association of Variable Star Observers* 28 (2000); Mattei and Kerriann H. Malatesta, "New Directions in Variable Star Research."

9. Payne-Gaposchkin, "The Dyer's Hand," 152.

10. Payne-Gaposchkin, "The Dyer's Hand," 203.

11. Payne-Gaposchkin, "The Dyer's Hand," 204.

12. Payne-Gaposchkin, "The Dyer's Hand," 225.

13. Payne-Gaposchkin, "The Dyer's Hand," 205.

14. Peggy A. Kidwell, "Cecilia Payne-Gaposchkin: Astronomy in the Family," in *Uneasy Careers and Intimate Lives*, ed. Pnina G. Abir-Am and Dorinda Outram (New Brunswick, NJ: Rutgers University Press, 1987), 232.

15. Gingerich, "The Most Brilliant Ph.D. Thesis," 3.

16. Harlow Shapley to Sergei Gaposchkin, April 21, 1938, UAV 630.22, box 33, Harvard College Observatory, Records of Director Harlow Shapley, 1921–1956, Harvard University Archives (hereafter Shapley Records).

17. Kidwell, "Cecilia Payne-Gaposchkin," 232.

18. Harlow Shapley to Sergei Gaposchkin, September 16, 1939, box 33, Shapley Records.

19. Shapley to Gaposchkin, September 16, 1939.

20. Harlow Shapley to Sergei Gaposchkin, April 28, 1938, box 33, Shapley Records.

21. Sergei Gaposchkin to Harlow Shapley, 1939, box 33, Shapley Records.

22. Sergei Gaposchkin to Harlow Shapley, March 15, 1940, box 33, Shapley Records.

23. Gaposchkin to Shapley, 1939.

24. Carol Dyhouse, *Girls Growing Up in Late Victorian and Edwardian England* (London: Routledge, 1981), 78.

25. Quoted in Dyhouse, *Girls Growing Up*, 36.

26. Katherine Haramundanis, "A Personal Recollection," in *Cecilia Payne-Gaposchkin*, ed. Haramundanis, 64.

27. Interview with Robin Catchpole, May 19, 2016.

28. Kidwell, "Cecilia Payne-Gaposchkin," 232.

29. Kidwell, "Cecilia Payne-Gaposchkin," 229.

30. Mattei and Malatesta, "New Directions in Variable Star Research," 70.

31. Hoffleit, "Reminiscences of Cecilia Payne-Gaposchkin," 87.

32. Letter from I. R. S. Broughton to Harlow Shapley, March 7, 1941, and Shapley to Broughton, March 14, 1941, quoted in Kidwell, "Cecilia Payne-Gaposchkin," 235.

33. Otto Struve to Henry Norris Russell, November 4, 1938, and Russell to Struve, November 7, 1938, quoted in Kidwell, "Cecilia Payne-Gaposchkin," 234.

34. Otto Struve to P. Buck, December 24, 1944, quoted in Kidwell, "Cecilia Payne-Gaposchkin," 234.

35. Henry Norris Russell to B. B. Cronkhite, October 28, 1948, quoted in Kidwell, "Cecilia Payne-Gaposchkin," 235.

36. Quoted in Kidwell, "Cecilia Payne-Gaposchkin," 235.

37. Haramundanis, "A Personal Recollection," 41.

38. Email correspondence with Katherine Haramundanis, August 2017.

39. Haramundanis, "A Personal Recollection," 42.

40. Gingerich, "The Most Brilliant Ph.D. Thesis," 4.

41. Haramundanis, "A Personal Recollection," 44–45.

42. Haramundanis, "A Personal Recollection," 40.

43. Payne-Gaposchkin, "The Dyer's Hand," 206.

44. Haramundanis, "A Personal Recollection," 61.

45. Payne-Gaposchkin, "The Dyer's Hand," 208.

46. Haramundanis, "A Personal Recollection," 40.

47. Haramundanis, "A Personal Recollection," 44.

48. Cecilia Payne-Gaposchkin to Otto Struve, April 13, 1944, quoted in Kidwell, "Cecilia Payne-Gaposchkin," 233.

49. Haramundanis, "A Personal Recollection," 44.

50. Payne-Gaposchkin, "The Dyer's Hand," 211.

51. Bart J. Bok, *Harlow Shapley, 1885–1972: A Biographical Memoir* (Washington, DC: National Academy of Sciences, 1978), 256.

52. Donald Menzel to Peter Gaposchkin, March 25, 1958, UAV 630.37, box 32, Menzel Papers, Records of the Harvard College Observatory, Harvard University Archives (hereafter Menzel Papers).

53. Cecilia Payne-Gaposchkin, "Note to Members of the Observatory Council," February 1958, UAV 630.37, box 32, Menzel Papers.

54. Menzel to Gaposchkin, March 25, 1958.

55. Donald Menzel to Peter Gaposchkin, March 10, 1958, UAV 630.37, box 32, Menzel Papers.

56. Haramundanis, "A Personal Recollection," 41.

57. Haramundanis, "A Personal Recollection," 41.

58. Payne-Gaposchkin, "The Dyer's Hand," 225.

59. Haramundanis, "A Personal Recollection," 60.

60. Helen Lefkowitz Horowitz, "It's Complicated: 375 Years of Women at Harvard," Lecture on History of Women at Harvard in Honor of Harvard's 375th Anniversary, April 23, 2012, https://www.radcliffe.harvard.edu/news/in-news/remarks-its-complicated-375-years-women-harvard.

61. Virginia Trimble, "Cecilia Payne-Gaposchkin: An Introduction," in *Cecilia Payne-Gaposchkin: An Autobiography and Other Recollections*, ed. Katherine Haramundanis, 2nd ed. (Cambridge: Cambridge University Press, 1996): xvii.

62. Payne-Gaposchkin, "The Dyer's Hand," 227.

63. Payne-Gaposchkin, "The Dyer's Hand," 225.

64. Haramundanis, "A Personal Recollection," 60.

65. Payne-Gaposchkin, "The Dyer's Hand," 211.

66. Haramundanis, "A Personal Recollection," 43.

Chapter 20

1. Katherine Haramundanis, "A Personal Recollection," in *Cecilia Payne-Gaposchkin: An Autobiography and Other Recollections*, ed. Katherine Haramundanis (Cambridge: Cambridge University Press, 1984), 64–65.

2. Haramundanis, "A Personal Recollection," 55–58, 62.

3. Haramundanis, "A Personal Recollection," 57.

4. Cecilia Payne-Gaposchkin, *Stars and Clusters* (Cambridge, MA: Harvard University Press, 1979).

5. Vera C. Rubin, "Cecilia Payne-Gaposchkin (1900–1979)," in *Out of the Shadows: Contributions of Twentieth-Century Women to Physics*, ed. Nina Byers and Gary Williams (Cambridge: Cambridge University Press, 2006), 167.

6. Owen Gingerich, "Cecilia Payne-Gaposchkin," *Quarterly Journal of the Royal Astronomical Society*, 23 (1982): 450.

7. Cecilia Payne-Gaposchkin, "No Wine So Wonderful as Thirst," *Radcliffe Quarterly* 41, no. 1 (1957): 11–12.

8. Cecilia Payne-Gaposchkin, "The Dyer's Hand: An Autobiography," in *Cecilia Payne-Gaposchkin*, ed. Haramundanis, 221.

9. *Harvard Crimson*, Nov. 6, 1970, 1.

10. Payne-Gaposchkin, "The Dyer's Hand," 233.

11. Payne-Gaposchkin, "The Dyer's Hand," 227.

12. Dorrit Hoffleit, "Reminiscences of Cecilia Payne-Gaposchkin (1900–1979)," in *The Starry Universe: The Cecilia Payne-Gaposchkin Centenary*, ed. A. G. Davis Philip and Rebecca A. Koopmann (Schenectady, NY: L. Davis Press, 2000), 89.

13. Jesse L. Greenstein, "An Introduction to 'The Dyer's Hand,'" in *Cecilia Payne-Gaposchkin*, ed. Haramundanis, 10.

14. Cecilia Payne-Gaposchkin, "Henry Norris Russell Prize Lecture of the American Astronomical Society—Fifty Years of Novae," *Astronomical Journal*, 82, no. 9 (1977): 665.

15. Payne-Gaposchkin, "The Dyer's Hand," 227.

Acknowledgments

First to Owen Gingerich. I visited this patient professor's office at Harvard's Astronomy Department so many times! Had he not suggested that Harvard University Press consider the manuscript, there would be no need to also thank:

A host of academic and publishing professionals, including Esther Newberg, my agent at ICM, and her assistant Alexandra Heimann. Patricia Mulcahy, a freelance editor whose expertise ranges from sentence doctor to structural strategist. Jeff Dean, my original editor at Harvard University Press (who early on wrote me an email with "green light" in the body of the message), and the rest of the HUP team: executive editor Janice Audet, editorial assistant Emeralde Jensen-Roberts, publicist Megan Posco, and especially senior editor Louise Robbins, for her meticulous fact-checking. Virginia Trimble, professor of physics and astronomy at the University of California, Irvine, and David H. DeVorkin, senior curator in the Space History Department of the National Air and Space Museum, who were so generous with their time and effort in sharpening the manuscript's scientific and historical accuracy. Madelyn Lugli, who made my archival research as easy as clicking on a Dropbox folder. Denise Bosco, a most resourceful photo researcher. Anne Thomson and Eve Lacey at Cambridge University's Newnham College. Mark Hurn at the Cambridge Observatory.

A wonderful group of family and friends, including Sylvia Auton, who read, and advised on, the first draft, chapter by chapter by chapter. Katherine Haramundanis, for her recollection of her mother's life and her collection of family photographs. Susan Gregory, who knew exactly what Cecilia had to put up with. Lanny Jones, former managing editor of *People Magazine*, who provided the spark.

And finally, Ann S. Moore, who bears the burden of reading before anyone else. She is saved for last on this list, but first in everything else.

Illustration Credits

Pages 129, 145: Photo courtesy AIP Emilio Segrè Visual Archives, Physics Today Collection

Page 141: HUP Fleming, Williamina (5b), Harvard University Archives

Pages 143, 208: Photo courtesy of AIP Emilio Segrè Visual Archives

Page 146: Photo by Margaret Harwood, courtesy of AIP Emilio Segrè Visual Archives, Physics Today Collection, Shapley Collection

Page 147: HUPSF Harvard College Observatory (19), Harvard University Archives

Pages 148, 149: Photo courtesy of Katherine Haramundanis

Page 155: Smithsonian Institution Archives. Image # SIA2007-0011

Page 156: HUV 1210 (2-2a), Harvard University Archives

Pages 159, 162: Center for Astrophysics | Harvard & Smithsonian, Photographic Glass Plate Collection

Page 164: Smithsonian Institution Archives. Image # SIA2009-1326

Pages 175, 181: AIP Emilio Segrè Visual Archives, W. F. Meggers Collection

Page 184: Courtesy of John G. Wolbach Library, Harvard College Observatory

Page 201: Courtesy MIT Museum

Page 252: Smith College Special Collections

Index

Note: Page numbers in italics indicate photographs.